THE
WIZARD
OF
QUARKS

THE
WIZARD

OF
QUARKS

A FANTASY OF PARTICLE PHYSICS

ROBERT GILMORE

COPERNICUS BOOKS
AN IMPRINT OF SPRINGER-VERLAG

Published in the United States by Copernicus Books,
an imprint of Springer-Verlag New York, Inc.
A member of BertelsmannSpringer Science+Business Media GmbH

Copernicus Books
37 East 7th Street
New York, NY 10003

www.copernicusbooks.com

Library of Congress Cataloging-in-Publication Data
Gilmore, R.S., 1938–
    The wizard of quarks : a fantasy of particle physics / Robert Gilmore.
      p. cm.
    Includes index.
    ISBN 0-387-95071-0 (alk. paper)
    1. Particles (Nuclear physics)—Popular works.   I. Title.
QC792.4.G55   2001
539.7′2—dc21   00-059029

Manufactured in the United States of America.
Printed on acid-free paper.

9 8 7 6 5 4 3 2 (Corrected second printing, 2001)

ISBN 0-387-95071-0   SPIN 10849692

To all my family, particularly the new arrivals.

CONTENTS

THE
WONDER
WORLD

It is sometimes said that the "cold light" of science destroys a sense of wonder—that you cannot find the world remarkable if you know that it is just atoms. Just atoms! This is like saying that you cannot appreciate the works of Shakespeare, the Bible, or for that matter the *National Enquirer,* because they are just words. It all depends on what's been done with the words or atoms.

Even if you feel that atoms themselves are cold, boring, and totally lacking in any interest (a view I would strongly contest), this is not to say that a world composed of atoms should be any less wonderful than one built from some sort of indeterminate, mysterious *stuff.* Consider a penny. There's not much you can do with it, is there? Then consider a billion pennies: ten million dollars. The subject has suddenly become more interesting. Finally, consider a thousand times this amount. Now the possibilities seem endless, limited only by your imagination.

Next consider not a penny but an atom. An atom on its own is a wonder, as I hope to show, but even if a single atom arouses no great interest, the possibilities multiply with many. Consider a number far greater than the quantity of one thousand billion that made pennies fascinating in the previous example. Consider a number some ten thousand million million times as great. Such numbers defy comprehension, but so many atoms can make many things. The atoms could combine to produce you or me, for example. They could construct any creature, artifact, or natural marvel that you could imagine, so long as it is roughly person-sized. Larger objects require even more atoms, of course. No longer can we say that the possibilities are limited only by your imagination. In addition to constituting anything you could possibly imagine, atoms could combine to produce all the enormously greater numbers of things that you could *not* imagine.

One of the messages of science is that the Universe is *not* restricted to what we can imagine. Some discoveries, such as quantum mechanics, are quite beyond our imaginings. There are more things in Heaven and Earth than are dreamt of in our philosophy, and it is among the wonders of science that it has shown us a few such things.

Science can only serve to increase our sense of wonder; it is "value added." It gives us new thoughts, new ideas at which to wonder. I cannot believe that a sense of wonder requires that we be ignorant and that wonder fades in the face of knowledge. If that is so, then it is a poor sort of wonder. Rather, understanding can amaze us and vastly enhance our sense of wonder.

I CAN'T UNDERSTAND IT!

It seems that people often do not accept that the findings of science should be wonderful and amazing. Occasionally people tell me they cannot understand what I am trying to tell them in my books. This may well be because I have not explained the material effectively. That is entirely possible, but I suspect that in many cases, it is because of their understanding of the word *understand*.

I think that what people frequently mean by *understand* is that when they have had something explained to them, they should realize that it "makes sense"—that it follows reasonably from what they already know and believe. You cannot expect this, however, when you are looking at something really new. What you discover will be strange, it will be surprising, and often it will be very hard to accept. The old saying that there is nothing new under the Sun simply isn't true. We do *not* already know everything.

When we examine a totally new situation, we may expect to discover totally new things. These discoveries will not necessarily "make sense" in terms of what we already know. They may even violate our "common sense," because we have no prior experience to go on. Common sense is a sort of distilled consensus of the cumulative experience of our lives, and in relatively common situations it is a reliable guide. But the quantum world is not within our normal experience, nor for that matter is the world described by the theory of relativity, in which objects are moving near the speed of light. Our intuition, derived as it is from our experience, does not help us here. We have no right to any preconceptions, and certainly we have no right to impose any preconditions about what Nature must be like. When we encounter something that is *totally* outside our previous experience, we must become like little children, learning about the world anew. Babies (as far as I can tell) have no preconceptions of the world; they must come to make sense of it afresh. By the same token, we must become "born again" quantum physicists.

We did not create the world, and Nature's rules are not ours to dictate. Whatever is, is. We can do nothing but discover it, attempt to describe it, and above all wonder at it. Wonder is the appropriate response, not disbelief or rigid determination to make it all fit into what we already know or believe.

A SCIENTIFIC APPROACH

I must admit that this book is not scientific. This is not because I try to impart a sense of wonder; that goes with the practice of science. Nor, I hasten to add, is it because the picture of the particle world that I have tried to paint is not one to which contemporary scientists subscribe; as far as I can manage, it is. What is missing is the experimental evidence. In this book I have used analogies to make many statements about the nature of particles, but I have given no evidence in support. Science, by contrast, depends on experiment. We have no innate knowledge of what the universe should be like and must discover by observation.

Many of the findings of science—and of quantum physics in particular—do sound totally implausible, but we must live with them because the account of the world that they provide is the only one that agrees with observation. This is the first and last requirement of science, and it is why scientists, who are as inclined to disagree among themselves as any other group of opinionated individuals, end up in agreement and all convinced of *the same thing*. Our models, then, must correspond to what we see, must be consistent with everything we see, and (where possible) should predict what we have not already looked at. This is not a case of one or two world-shattering experiments. That is not how science works. The idea of one "great experiment" is undesirable (though sometimes cost limits the number of observations, as it does in particle physics). Most of physics is supported not by just one observation, but by thousands—and more probably millions—of observations made year in and year out. These observations are recorded by a whole army of experimenters, including the scientists in great national laboratories, researchers in universities, and teachers and students in high school laboratories. The accepted physical picture of the world has to agree with the measurements they all make. It has to agree with every single one of these measurements, and of course it must also agree with what we all observe in our everyday lives. Often the theories can correctly predict the results of observations before they are made. The sheer overwhelming weight of all this evidence makes it very difficult to entertain alternative views.

It has been said that experiencing the wonder of science is like feasting in a gorgeous banquet hall that can be reached only through a long and dreary kitchen. The kitchen represents the process of performing and interpreting experiments. This process can be complicated and tedious, but experimentation alone gives us cause to believe in the conclusions of science. In this book, I invite you into that hall and give you a taste of the wonders in the banquet, but it is important to remember that it is only by visiting the kitchen that you can tell if the banquet is safe to eat—if the wonders are actually true. We owe all the delicacies we will savor to the painstaking scientific work that has taken place—and continues even as we dine—in the kitchen.

. . . AND THERE IS WONDER APLENTY

There is plenty in the current picture of the atomic and subatomic world to evoke surprise and wonder. To wit . . .

1. There is the prevalent effect of *interference* (Chapter 2). At the atomic level, the large-scale concepts of particle and wave cease to be appropriate. Instead there is *amplitude,* a thing that does not mesh well with our everyday experience. In this world there are no choices. There is an amplitude present for everything that *could* happen, and the amplitudes all interfere. They add *and subtract,* so it is the total collection of amplitudes that determines what you are likely to observe.

2. There is the ever-present role of *phase* (Chapter 2). Classically, phase is a "wave-like" thing. It describes some sort of relative orientation of the different amplitudes and determines how they affect one another, whether they add or subtract or reach some sort of intermediate compromise. Phase is vital.

3. There are *quantum fluctuations* (Chapter 3), as described by the Uncertainty Principle. Many quantities—in particular, momentum and energy—may vary a little. The total energy is constant over the long term, but it fluctuates over momentary periods.

4. *Virtual particles* (Chapter 8) are born of energy fluctuations. The shorter the period over which energy is considered, the more it can vary. Over very short periods, it may vary enough to create the rest-mass energy of new particles. Such virtual particles may be born from nothing and to nothing return, but in their transient existence they are responsible for all the interactions that hold our world together.

5. The Pauli Exclusion Principle (Chapter 4) arises from the distinction between particles whose amplitudes change sign when two such particles are exchanged and particles whose amplitudes do not do so. Does this seem like an academic point that should rank in relevance well below the medieval debate about how many angels could dance on the head of a pin? On the contrary, it has interesting consequences, such as the existence of atoms and the fact that you do not sink into the ground.

6. *Renormalization* (Chapter 6) is an attempt to deal with the infinite energy that should accompany the infinite set of virtual particles allowed by point 1 in this list. In effect, the particles' energy is buried in an infinitely deep hole, with just its observed rest-mass energy poking above the surface.

. . . and all that comes before we even get to *quarks*.

I invite you to come in, seat yourself comfortably, and partake of the feast that is a fundamental understanding of the world we all inhabit.

ACKNOWLEDGEMENTS

I wish to thank the CERN laboratory for permission to use pictures from their web page as background material in some of my illustrations.

I also should like to point out that the purely mythical Kingdom of Cern, which I invented to discuss the proliferation of "elementary particles" in the middle of the twentieth century, bears little resemblance to the real CERN laboratory, now confidently tackling the physics of the twenty-first.

There are very few centaurs in CERN.

UPDATE AND FURTHER READING

Robert Gilmore has written two other books in a similar vein, namely *Alice in Quantumland* and *Scrooge's Cryptic Carol*. Both books are published by Copernicus.

There do not seem to be very many new books on particle physics for a general readership. Current information is best found on the Internet. These sites will change from time to time, so my intention is to list a few current ones on my own website

www.phy.bris.ac.uk/allegory.

Shortly after a book is published I tend to think of things I wish I had included. Any such afterthoughts and other material that I feel I wish to share will also be included in my web site.

CHAPTER 1

THE

WITCH

OF

MASS

The subway car gave a sudden lurch.

That was not in itself surprising, for Dorothy had found the ride to be pretty bumpy on the whole. Her Uncle Henry and Aunt Em had had a particularly good corn harvest that year and had decided to celebrate with a trip to the Big City. So here they were, and, naturally, they were taking a subway ride. No sooner had they settled into their seats than they discovered that the two people opposite them were from Kansas too, and soon all four adults were deep in a discussion of corn prices.

Dearly as Dorothy loved her aunt and uncle, she had long ago begun to feel that there must be more to life than the price of corn. She soon became bored and moved off to investigate the train. This did not take her very long; it was a quiet time of day, the train was not crowded, and one partly empty subway car looks very much like the last. Eventually she found a car that was completely empty, took one of the vacant seats, and sat looking out the window. There was not, it must be confessed, a great deal for her to see there either. Intriguing cables hugged the tunnel walls, and now and then a lighted station framed the bustle of people getting on and off. Once in a while, the tedious length of tunnel was relieved by a short side passage lined with pipes and cables and ending at a concrete wall, before which a ladder rose to unseen regions beyond. Dorothy wondered what might be at the top of the ladder, but there was no way to know, and each side passage vanished quickly from her sight. Mostly what she saw out of the window was darkness and her own reflection. Then came that sudden lurch.

Up to that point the train had shaken and rattled as it rolled over the rails, but its occasional jolts and lurches had been nothing like this. They had been your ordinary everyday sort of lurch, not a slippery, squirmy, "you have just tripped over a discontinuity in the space–time continuum and had better readjust your perception of the Universe" sort of lurch.

This was one of *those* lurches.

Afterwards the train continued to run quite smoothly—a great deal more smoothly than before, to be honest. The walls of the tunnel seemed smoother too, though it was still dark outside, so it was hard for Dorothy to be sure. The train passed another side tunnel, and as Dorothy glanced down it, she noted a difference from any side passage she had seen previously. Before, the tunnels had been lined with cables, and a ladder had been visible at the end. *This* tunnel had smooth, featureless sides and ended in a bright circle of light with a blurred outline. It reminded her of the time she had peered down a microscope at school. Through this circle of light she seemed to be looking down on several weathered buildings, all set in a vast sea of waving corn. Above one of the buildings there proudly projected a weather vane in the shape of a running horse. Despite the unusual overhead viewpoint, Dorothy realized that, surprisingly, she was looking at her uncle's farm: her own home.

No more than a glimpse and the vision was whisked away by the motion of the train as it rushed past the end of the tunnel. After a few moments of darkness outside the window, she saw another tunnel. It was identical in all respects to the one before, except that the circular lens at the end was now focused on the meticulously scrubbed top of a wooden kitchen table. Beyond the table lay what looked very much like the kitchen in her own home back in Kansas. The sugar bowl had been overturned, and Dorothy saw, marching determinedly to and fro across the table, a line of specks that she recognized to be a trail of foraging ants.

Once again the train rushed away from the opening, but soon there came another. At the end of this tunnel, Dorothy had just time to see a great armored shape with spreading, angular legs. The antenna-crowned head swung around and seemed to be gazing at her with great compound eyes on either side as she passed out of sight.

Another tunnel and another view came into sight. In this she saw what appeared to be an endless array of hazy spheres, like a huge number of out-of-focus beach balls joined together in two thick ropes that twined about one another. She could make out no detail, but it was obvious that each and every sphere that formed a part of this double helix was in motion, vibrating about its proper position in the array. Some vibrated violently, others more gently, but all were in motion, and soon all were left behind as the train moved on.

The next view was filled with a vague, fuzzy shape. Dorothy somehow had the impression that the object she saw was quite complicated, but her perception of it was very *uncertain*.

Dorothy was entranced by this succession of fleeting peep shows. She had no idea that the Subway Company provided this sort of entertainment for its passengers. Within the car the light was dim and grey, and of course outside the window it was dark most of the time, so the pictures came as a welcome relief. Abruptly, Dorothy realized that the persistent rattle of the wheels had stopped, and then came a sensation of falling. The view through the windows was no longer black, but a foggy and steadily lightening grey, as though she and the car were wrapped inside a cloud that was beginning to disperse.

Then came a shock so sudden and severe that if she had not been securely seated, she might have been hurt. The fall was over. The door of the car slid back, and the fog quickly evaporated to reveal a landscape of marvelous beauty. There were lovely patches of lawn and banks of gorgeous flowers on every hand—all very different from the dry grey stone of the city. All was strange and brilliant to behold, though beneath it all, she still had a feeling that even this gorgeous landscape was in some way, deep down, colorless.[1]

---

[1] The role of *color* in particle physics is discussed in Chapter 9.

Dorothy ran out into the meadow and looked around. Her gaze traveled over meadows and woods and flowers. She looked about in wonder and slowly turned around, marveling at everything she saw. As she turned, her gaze suddenly encountered a grossly incongruous element. Squatting alone and isolated in the midst of this rural idyll was a harsh, alien shape: a single subway car, drab and covered with graffiti and looking utterly out of place in its present surroundings. As she looked at this unpleasant sight, Dorothy was horrified to observe, protruding from beneath its wheels, a pair of stout legs clad in striped stockings. No sooner had she sighted them than they withdrew under the carriage, and then the whole massive vehicle *lifted slightly*. From beneath it a figure came crawling, floating, sliding—there is really no word that will adequately describe the way this person moved from beneath the car. She came to a stop, suspended in the air before Dorothy, with the demeanor of someone who is perfectly familiar with the *notion* that things normally tend to fall down and lie upon the ground, but who simply doesn't believe such notions apply to *her*. Her position was the more subtly unnerving because her body, the fall of her hair, the drape of her robes—everything about her—was aligned at a slight but definite angle to the vertical. Now that Dorothy could see her plainly, she was revealed as a woman of ample, not to say enormous, dimensions. She looked at Dorothy as severely as was allowed by features nearly lost in her great round face.

"Ah, my child. Were you the one responsible for dropping this *equipage* on top of me?"

"Oh, no, Ma'am. At least I don't think so. I was just riding in the subway, you see. . . ."

"Well never mind, there's no harm done. But I shouldn't make a habit of dropping these subway things on people. Someone else might get hurt, though of course in my case there was no danger of that."

"And who *are* you, Ma'am, if I may ask?" inquired Dorothy, with a little curtsy just to be on the safe side.

"Oh, I am the Witch of Mass, the personification of the gravitational interaction. That is another name for the force of gravity, the force that holds the planets in their orbits and keeps you and your kind bound to the surface of the Earth. You may call me G, if you wish. That is how I am often known."

"Are you really a witch?" asked Dorothy, choosing to tackle the simpler concept first. "I do not think I have ever met a witch before."

"You will have met *me,* or rather you will have experienced the effects of my interaction of gravity. There are four of us witches in all, and each of us represents some form of force or interaction experienced by matter. You will in one way or another have seen the effects of all these interactions, though some of them may have been very far from evident."

"Excuse me," said Dorothy, feeling rather at a loss, "but what do you mean by an interaction?"

"An interaction is the way in which one particle, one isolated fragment of the whole body of the Cosmos, can have any effect whatever on another of its companions. You see, the whole world as made up of different materials—solids, liquids, and gasses—that surround you in incredible profusion. Each may appear to you to be a thing in itself, completely uniform and continuous in its composition, but it is not. None of them are. Everything you see, everything that makes up the world around you and the great universe beyond it, even your own body: All are made up from great numbers of tiny particles. You are not aware of this because the numbers are indeed great and the particles much too tiny for you to see. All the various materials are combinations of these *basic building blocks* of the universe, and they so combine because of the *interactions* between the particles. Were it not for interactions, there would be no feature or structure in the whole Universe. There would be only a formless chaos of isolated particles, each doomed forever to total loneliness and with no awareness of its fellows, even though it is in their swarming midst."

Here the Witch of Mass paused for breath, clearly proud of her key role in the Cosmos. She settled toward the ground and moved closer to Dorothy. Two large rocks that happened to be in her path moved obligingly out of her way.

"I am not the only witch here. As I mentioned, there are four interactions and so there are four witches to represent them. Of all my fellow interactions I am both the greatest and the least. I am greatest because the scope of my influence is extensive beyond all imagining. It is through me that the Earth and all the planets are guided in their annual dance around the Sun. Further yet, I persuade great carpets of stars to crowd together within swirling galaxies and even cause groups of galaxies themselves to fall together into tremendous clusters. Though my concern is mostly with the great patterns of the Cosmos I have an effect on a more local scale as well. Not a sparrow falls but gravity is there. Thus I am the greatest of the interactions in that my influence pervades the entire Universe, but I am also least because the effect I produce is weak. Very weak," she added rather ruefully. "Compared with the interactions made possible by my sisters, the effect that one particle has on another through gravity alone is microscopic. Indeed the word *microscopic* does not do justice to the disparity."

The stout figure of the Witch settled gradually to the ground and looked around her, rather cautiously Dorothy thought.

"One of my sister interactions is the Witch of Charge, the personification of the electromagnetic interaction. Her effects are observed most clearly in electrical processes and in magnetic fields, but she also works behind the scenes in everything you see around you. She is called EM."

"Oh!" interrupted Dorothy enthusiastically. "I have an Aunt Em, you know."

"I shouldn't think there is any connection between EM and your aunt, child," said the Witch, peering down past her massive bosom at Dorothy through a tiny pair of eyeglasses. "EM represents one of the four great interactions that control the world. Her reach is inherently as great as mine, and her strength is greater by an amount you cannot begin to imagine. If you consider two particles that have both mass and electric charge, then the force between them due to gravity is *much* less than that due to their electric charges. I speak not of a million times weaker, or a million million, or a million million million. . . ."

Here G indicated a beam of sunlight that slanted through the branches of a tree overhead. In its brightness A myriad of tiny dust motes sparkled "My strength compared to EM's is less than the size of the smallest speck of dust that you can see glinting in that beam of sunlight is compared to the distance at which the most powerful telescopes can see stars in the far depths of the Universe." The Witch gestured vaguely at the brilliant blue sky overhead. Dorothy saw no sign of stars, but she was prepared to believe that they were there. "Less indeed," continued the Witch emphatically, "than that tiny dust mote is in comparison to the whole distance that light could possibly have traveled since the Universe began."[2]

"That sounds pretty tiny," agreed Dorothy, "but if gravity is as weak as all that, how can it be that gravity, and not this other, enormously powerful interaction, guides the planets?"

"Cooperation, my dear!" replied the self-styled Witch of Mass with a satisfied smile, leaning over and speaking confidentially into Dorothy's ear. Dorothy thought it was rather like having a *tête-à-tête* with a mountain. "Teamwork is the answer. When particles feel the pull of gravity, they all work together to the same end. Every single one attracts every other one, without exception. The effect of gravity as experienced by any two particles on their own may be infinitesimal, but when every particle in the Universe is attracting every other, then on the grand scale this overcomes all competition. There are a *lot* of particles in the Universe!"

The Witch straightened up again, still looking enormously satisfied with herself. (Dorothy felt that *enormously* was exactly the right word.)

"EM's strength is inconceivably greater than mine," she continued, "but in spite of that, you are less aware of electrical attraction in your world than you are of gravity. This is because EM is always at war with herself. She is the Witch of Charge, and electric charge comes in two types: positive and negative. It works in such a way that opposite charges (a positive and a negative) will attract each other, but two like charges (two positive or two negative) will *repel*. They will push one another away, whereas my gravity would never reject *any* particle. The Universe contains both positive and negative charges, and positive and negative are present in equal number. The attraction that the charges of opposite type feel for each other is balanced by the repulsion between charges of the same type. The balance is so precise that the huge forces of EM's electrical interaction are quite hidden," the Witch continued rather smugly.

"In some instances the effects do not quite balance, but in general the balance is exact, or as close to exact as you could possibly imagine.[3] So you

---

[2] Honest! In fact this comparison does not do justice to the discrepancy. The comparison depends on which particles you choose, but for two protons the strength of gravity is down from that of electromagnetism by a factor of about 10^{36}. If you assume the smallest dust particle you might see is a thousandth of a millimeter across, which is almost certainly too small, then multiplying by the above factor gives a distance of more than *one hundred million mega light years*. This is a long way, more than a thousand times as far as light could travel in the lifetime of the universe.

[3] It is difficult to comprehend just how *precisely* the positive and negative charges in the world must balance each other. Despite the inherently vast strength of electromagnetism, in practice its effects are so reduced that gravity dominates completely on the large scale.

see, even though she is much stronger than I, her strength is generally wasted in a struggle with herself, and my cooperative approach ends up by dominating. Her internal struggle is so exquisitely balanced that the overall effect of her huge electrical forces is not felt, and my gravity reigns supreme. It dominates on the large scale, that is," qualified the Witch. "You may have noticed that the effect that gravity has on something becomes less the smaller that something is. A fly can walk up a wall against the pull of gravity with no difficulty, whereas an elephant would have very little success. A really large animal, such as a whale, cannot even bear its own weight with comfort unless it is supported by water. The smaller the object, the less it is affected by gravity, because there are fewer particles within it to cooperate with the interaction. For the atoms *within* a material, the relative effect of gravity is much less again."

With these words the Witch rose and floated (if that is the best way to describe the movement of someone who is simply rejecting any effect of gravity) across to the subway car. Possibly as a concession to Dorothy's sense of propriety, she settled again to the ground and stood beside the blatant lump of grey solidity. She made a slight gesture, and the mass of metal stirred as it had before and lifted slightly into the air. Another gesture and it rose further, accelerating ever faster until it was so far overhead that it looked like a little toy train, spinning slowly in the blue sky.

Then Dorothy noticed that it looked perhaps a little less tiny. Then without question it began to grow, and she knew that it was falling, rushing down at her out of the sky. The ground shook as it crashed down a little way from where she stood. There was the scream of crumpled metal, and a great cloud of dust flew up to obscure the carriage. As the dust settled, Dorothy could see that the subway car, never an object of great beauty, was now a shattered wreck.

"There you see the effect that gravity can have. The aftermath of a serious fall can appear devastating, but if you look more carefully, you will see that this object is not so greatly changed. In many places it is twisted and distorted, certainly, but it is largely recognizable. The wheels are still wheels, though no longer squarely mounted on their axles. The seats are still seats, though torn from their fastenings." She gesticulated at the contorted fragments that lay around. "More significantly, although the shape and constuction of its parts may have changed beyond recognition, the essential nature of the *materials* from which they are made has not changed. The metal panels may be twisted, but they are still metal. The glass windows may be shattered, but they are still glass. The arrangement of the particles within the materials—the atoms that make up the metal or the glass—has not been altered. It is on this scale, the scale of the atom, that EM reigns supreme."

The Witch paused dramatically to emphasize her point and then continued. "As you look at matter in finer detail, considering its internal structure and even the nature of the atoms that compose it, the balance of EM's forces becomes less exact. Because the positive and negative electric charges are not *quite* in the same place, there is a slight imbalance in the electrical forces they produce, and even this small residual force is so relatively huge that you may neglect gravity in comparison. The atom—any atom—is a creation of electrical effects alone.

"I do not know much about the small scale," continued the Witch. (Dorothy could well believe this. The Witch certainly did not look as though she would know much about anything *small*.) "Atoms and particles and all that sort of thing are not really my concern. I do not trouble myself to try and tell one from the other. They all have some mass and energy, so they all feel the effect of gravity, however weakly, but their particular natures do not concern me. If you explore even smaller sizes, you will discover further particles, I am told, all sorts of them. Then you may encounter my two other sister witches, who may appear to you to have little influence on any wider scale."

"First is the Witch of Color, the spirit of what is called the strong interaction. Most people find her a little difficult. She comes on a bit strong, you might say. Her full name, Quantum Chromo Dynamics, is rather a mouthful, and she is usually known by her initials, QCD. Second, there is the Weak Witch, personification of the weak interaction. You want to watch out for her, because she's sneaky. She creeps up on particles and *changes* them. Neither of these last two has very obvious effects on the larger scale, the scale of the world that you inhabit."

"Speaking of that world," Dorothy said carefully, "do you by any chance know how I might get back to it?"

"I am afraid that is not for me to say. I take the broad view of things. As I have told you, the world of matter—the world of atoms and particles in general—is not really my concern."

"Do I need to know about particles to get home again then? Can you tell me anything about them?" asked Dorothy anxiously.

"Haven't you been listening?" retorted the Witch. "No, I cannot! They are all one to me. If you want any detail, you should go and visit the Wizard of Quarks. Particles are his concern. If you want to get home, you must visit the Wizard and ask his advice."

"Where might I find him?" asked Dorothy.

"Why, in his palace in the Emerald (and Ruby and Sapphire) City," replied the Witch. "I'm sure he can tell you all you want to know, and probably a great deal more besides."

THE FOUR BASIC INTERACTIONS

There are usually held to be four basic interactions (or forces) that act on and between particles. These may be, in a sense, different aspects of one *unified* interaction and they may become indistinguishable at high enough energies (see Chapter 13). For practical purposes, however, they are four distinct interactions.

1. **Gravity:** This is an interaction between all particles that have mass. This means all particles with energy, because energy and mass are the same thing (Chapter 6). It is a long-range force, falling off in intensity as the square of the distance between the particles concerned. At normal energies, the force of gravity between two particles is extremely weak and cannot in practice be detected.

2. **Electromagnetism:** This is an interaction between particles with electric charge. Like gravity, it is a long-range force, but between two particles such as protons, it is phenomenally stronger. There are two signs of charge, positive and negative. Charges of the opposite sign attract; those of the same sign repel each other. This results in a balance of electrical forces so exact that the much weaker force of gravity dominates on the large scale.

3. **The strong (color) interaction:** This is an interaction between particles with color charge. The only such particles we know are the quarks. Leptons such as the electron do not feel the color interaction at all. This is similar to the way electrically uncharged particles do not experience electrical forces. The strong interaction results in a very tight binding on the small scale and can resist the electrical repulsion that exists between protons when they are crowded close together within an atomic nucleus. It is effective only over a short range (see Chapter 9).

4. **The weak interaction:** The weak interaction has a shorter range and is much weaker than the strong interaction. Despite this, it was observed because it does things that other interactions cannot do. For example, it can convert a neutron into a proton together with an electron and an antineutrino. One particle turns into three different particles, as seen in nuclear β decay (see Chapter 13).

"But how do I get to the Emerald (and Ruby and Sapphire) City?" persisted Dorothy plaintively. She had difficulty in pronouncing the parentheses, but thought she did it rather well.

"Isn't it obvious?" The Witch's patience was wearing thin. "I have told you already that whereas interactions provide all the coherence of the Universe, particles provide the building blocks from which it is made. To find the Wizard of Quarks you must follow the *Building Block Road.*"

The Witch of Mass gestured toward a gap between the trees, and Dorothy could see beyond, set into the ground, four stone slabs on which were carved the words

EARTH

AIR   FIRE

WATER

From between these slabs a paved road came spiraling out and ran off into the distance. On each slab was carved a name or symbol. The letter **e**

THE WITCH OF MASS

predominated, but a few bore the symbols **p** or **n**. Both α and Ω lay upon this road, but they came farther along the way.

By the side of the road Dorothy saw stones set upright in the ground, like milestones, each carved into the form of a letter h with a bar across the top, $\hbar$. Those nearest to her seemed very small, but they became larger as the path progressed. "What are those?" she wondered aloud.

Surprisingly enough, the Witch seemed to understand perfectly what she meant. "Those are indicators of Planck's Constant, the universal constant that marks the scale of quantum behavior. Universal constants," she continued before Dorothy could ask, "are quantities that have a single absolute, definite value invariable throughout space and time. They are the fixed points in a somewhat variable Universe."

PLANCK'S CONSTANT, $\hbar$

Planck's constant is at the very heart of quantum physics. It is one of a very select group of *fundamental constants* that also includes the basic unit of electrical charge e and the limiting velocity c (often called the velocity of light). This select band of values seems to be basic to the very structure and nature of the Universe we inhabit.

Planck's constant sets the scale on which quantum effects are dominant. We are usually unaware of them because we live on such a coarse scale that $\hbar$ is too tiny to notice.

The size of $\hbar$ gives the *granularity* of the world, the scale on which the world separates into myriad little separate lumps such as atoms. The value of $\hbar$ determines the size of atoms and the discrete packets of light called photons. Planck first introduced the notion of light coming in little packets, or quanta, in a rather subtle argument about the emission of radiation from hot objects. Einstein made the idea much sharper in his theory of the *photoelectric effect,* the emission of electrons when light falls on a metal surface.

"But," protested Dorothy, "if it is as unchanging as you say, how is it that the markers I see are getting bigger and bigger as the path progresses? That does not look very constant to me."

"Appearances can be deceiving," replied the Witch. "It is easy to be misled about the relative sizes of distant objects. You see the road markers as increasing in size, but perhaps it is the road that is becoming steadily smaller. This path will carry you down to a scale on which quantum effects will become important. In fact, they will be dominant."

"But what are quantum effects?" asked Dorothy desperately.

"That you may discover for yourself as you travel along the path. Off you go then!"

With those curt parting words, the Witch turned away. Dorothy politely thanked her for her instruction, but that gigantic personality seemed to have lost interest. Dorothy could see no reason to delay further, so she set out resolutely along the paved way.

As she walked, she examined the roadside markers that they had just been discussing. The first few she passed were tiny—scarcely visible above the grass. In fact, it occurred to her, there might well have been any number of them further back along the road, too small for her to notice. As she made her way along, firm in her intention to find out something about these particles that were apparently so basic to everything she knew, the roadside stones became steadily larger until they reached the height of a normal road sign or greater.

As she marched ahead, she seemed to hear voices from all around her, saying in different tones,

> *"Follow the Building Block Road,*
> **"Follow the Building Block Road,**
> *"Follow the Building Block Road,*
> **"Follow, follow, follow, follow,**
> **"Follow the Building Block Road. . . ."**

CHAPTER 2

THE
OBSERVANT
SCARECROW

As Dorothy walked down the road, she noticed that the carved stones she passed at the roadside continued to increase in size until finally they came up to her waist. The sequence ended abruptly when the stones had reached this substantial size, and as Dorothy continued on her way, she soon left the last of the stones behind.

After marching along for a short time, she found herself passing a meadow on her left and paused to admire the view. She found it most unlike the many meadows she had known in the past. Anyone from her part of the world was something of an expert in fields and meadows and prairies in general, but this was different. There was not so much a bright golden haze on this meadow, but rather the meadow *itself* was hazy. Some way off, she could see figures moving around and reaching from time to time within the haze. Most times they appeared to come up empty-handed, but sometimes one or another would drag out *something* from the general mist. Dorothy peered more intently ahead to try to discover what was happening.

"They are rounding up electrons," said a voice from just above her head.

Dorothy looked around in surprise, but could see nothing other than a post planted securely in the field behind her. Glancing upward, however, she saw a scarecrow dangling from the top of the pole and staring out over the meadow with wide eyes painted on the small, stuffed sack that made up his head. He had on an old pointed hat, and heavy boots dangled from below his trouser legs. Both arms hung limply at his sides, ending in a pair of old work gloves, and untidy tufts of straw escaped from the ends of his shirt sleeves and

trousers. As Dorothy gazed at him in puzzlement, he suddenly nodded his head to
her in a friendly way.

"Did you speak just now?" asked the
girl in wonder.

"Certainly," answered the scarecrow.
"How do you do?"

"I'm pretty well, thank you," replied
Dorothy. "Would you please repeat what
you were saying about electrons?"

"Certainly, but would you first get me
down? After spending my entire life of
duty in this field, I think I am entitled to
leave my, er, post for a bit." Here he
chuckled at his own witticism.

Dorothy carefully lifted him down.
He was larger than she, so she feared the
task might be rather awkward for her, but
being just stuffed clothing, he was very
light. She set him down gently so that he

R Gilmore

stood upright before her, with only a slight tendency for his legs to fold up
beneath him. His arms still hung limply at his sides. (As she came to know
him better, she was to discover that his arms and legs always tended to dan-
gle unless he was using them deliberately for some purpose.)

"Right!" continued the Scarecrow as he swayed vaguely to and fro. "You
were asking about the electrons. As you may have realized by now, what you
probably thought of as a meadow is in reality a cloud or collection of many
electrons. This is usually called an amplitude. You are seeing a many-electron
amplitude as clearly as you are able. At the moment, some of the electrons
are being collected for use elsewhere."

"No, I certainly did not realize that!" returned Dorothy. "I do not know
anything about such things, and I do not see how you possibly can either.
The scarecrows on my uncle's farm do not seem to know much about anything, if you don't mind my saying so."

"Well, you might ask," the straw-stuffed man replied. "It is true, as you
are implying, that I have no brains, but I do keep my eyes open. I pretty well
have to, don't I? They have been painted wide open." Another chuckle. "In
any case, I see a lot. I observe everything that happens in my vicinity. So although my head may not contain any brains, it is absolutely stuffed with *information*—with observations that I have made."

"Oh, I see," said the girl, though she did not feel totally convinced. "Well, I am afraid that I did not at all understand what you said just now. What do you mean by an amplitude?"

"Why, an amplitude is the essence of any particle or other simple physical object. The amplitude says as much about the object it describes as you can possibly know about it. It *is* the object, as far as you are able to perceive it. So all that you know of particles such as electrons is their amplitude."

"Why does it all seem so fuzzy then?" asked Dorothy in some confusion. "I thought that particles were little, hard, sharp things, like marbles shrunk to tiny dots."

"I think that the best answer I can give to that," replied the Scarecrow, "is *no!* On the small scale, particles are not really like that at all. Things may seem sharp and precise to you in your large-scale everyday world, but once you are down on the level of Planck's Constant, things have a very different aspect. Come and see for yourself." So saying, the Scarecrow marched over to a nearby swelling in the fuzzy mist that covered the area. *Marched* is probably the best description of the way he moved, though it was unlike any marching that Dorothy had ever seen before. He would raise one foot at a time high in the air and swing it forward, while the rest of his body hung limp and inert. Whatever the right word was to describe it, it worked. Soon they were standing by a great swelling mound that rose from the overall foggy covering.

"The meadow has been planted unevenly, and the density of the electron cloud is very irregular. The *probability distribution* for the electron amplitudes varies from place to place."

"What is a probability distribution?" asked Dorothy.

Her straw-stuffed friend answered, as she feared he might, "It is a distribution of probability, the probability that you will find an electron at a given position."

"But where does probability come into it? Surely things in science are definite and predictable. In general, anything must be definitely in one place and be heading definitely in the direction of another."

"So you might think," replied the Scarecrow. "You might believe that nature is completely deterministic and predictable *in principle.* But even if that were so, there is no conceivable way that anyone could ever *actually* predict the motion of each tiny atom in a vast collection of atoms that are all colliding and bouncing off one another. You cannot readily observe a single atom accurately enough, let alone the vast numbers encountered in practice. All you can do is speak of a probability distribution for the atoms. You may estimate the probability that atoms will travel in a given direction and how many will go fast or slow. You can say nothing about individual atoms."

The Scarecrow paused so that Dorothy could consider the paradox of how things might be predictable in principle, but still could in no way ever be predicted.

PROBABILITY

Probability is not a novel feature of quantum physics alone. Even in classical physics, you can describe the motion of atoms only in terms of probability distributions. For example, you can never calculate the motion of all the atoms in a gas. You can only speak of the probable distribution of their velocities and directions of motion.

What is remarkable in quantum physics is not the use of probability distributions *per se,* but the way in which they derive from *amplitudes* that have phases (see below).

"Here, in a quantum state, every amplitude has a probability distribution that goes with it," he went on. And before Dorothy could ask a further question, he continued quickly. "The amplitude gives the best description you can have of what actually *is*, but this is not at all the same as what you will find if you actually *look*. One of the first things you must know is that there is no such thing as *just looking.* Let me tell you that I have had plenty of experience in observing by now, and I know that any way in which you can make an observation must interact with *and will change* the thing observed. There is no way that you can observe something without some physical interaction. If you look at it, that means a photon of light must have bounced off it. Such interactions will in general change what you are looking at."

Mesmerized by his own discourse, the Scarecrow continued. "A probability distribution tells you the probability of finding one condition or another when you do look. It tells you, in effect, the relative amounts of what is there to *be* found. Whenever you do look, you will observe one or another of the possible options, and when you make such an observation, why then the amplitude actually changes. It changes to an amplitude that allows only *one* possible result, the one that you have just seen. There is no longer any uncertainty at that stage, because you *have* seen what you have seen. As time goes by, however, you will usually find that because you are no longer watching or interacting with the system, the amplitude will steadily grow to include other possibilities as well."

"Anyway," he concluded abruptly, "there is a local maximum in the amplitude at this place, so you are more likely to be able to find an electron here. Just reach in, feel about, and see what you can find."

Dorothy obligingly reached out into the surrounding bright fog. As she did so, she was surprised and horrified to see her own hand become correspondingly fuzzy as soon as it entered the fog. There, extending from the sleeve of her dress, was her own arm as she had always known it, but below the elbow, it spread out and became blurred and indistinct.

R Gilmore

"What has happened?" she cried out in alarm. "What has become of my hand?" She snatched it back hastily and was relieved to see that it again looked as it always had.

"Why, it entered into the fog of reality and, as a consequence, became more real."

"More real!" echoed Dorothy. "It didn't look at all real to me. It was like a dream or a nightmare. Now it looks perfectly real and back to normal."

"Ah, but that is just an *illusion of normality*. On the scale of being that you now inhabit, a scale where Planck's Constant can no longer be ignored, such a fog *is* the reality. Your notion of normality, with everything sharp and precise, *is* an illusion. It is an illusion that comes from your being unable to see the detail of your own world precisely enough.

"Here, everything is *quantized*," asserted the Scarecrow emphatically. "We are now on a scale where we deal with atoms themselves, with the simple basic parts that in your world are too tiny for you to notice. Something as complex as your normal body contains unimaginable numbers of atoms and could not possibly exist on this scale. However, for your own peace of mind,[1] you experience the illusion that you and the folk you meet are much as they would be in a large-scale world. This illusion persists as long as you do not make too close a contact with the reality around you."

The Scarecrow paused to let his words sink in. "The world loses its sharp precision on the scale of Planck's Constant. It is like a newspaper photograph, which appears clear until you look at it closely, whereupon you find it is made up of many little dots, and the sharp picture becomes blurred."

---

[1] It also avoids my having to make drawings in which *everything* is simple and fuzzy and quantized.

"Just a minute," interrupted Dorothy. "You said that you had spent your life here in this field, so how do you know anything about newspaper photographs?"

"Oh, I have observed plenty of them," the Scarecrow lamented. "It is quite disheartening how much trash folk will dump in the countryside." He sighed wistfully.

"Here, however," he resumed, "the dots that make up the big picture are the basic elements of which the world is composed. This is just as true where you came from, but because the 'dots' are on a scale given by Planck's Constant, $\hbar$, they are far too small for you to perceive them in your usual existence. If you look closely enough into the fabric of matter, there is a blurring of position and movement, of being and becoming, that cannot be resolved any more precisely than this. If such indefiniteness as you see on this scale seems unreasonable, it is partly because of your preconceptions of how position in space should be. An electron, for example, is a definite object with its own quite specific character. It is just that its characteristics do not happen to include a single, unique position in space."

"Why should it be like that?" asked Dorothy, who was vaguely offended by the idea that things were not as they had always seemed. "It doesn't seem right. Why does the world have this deep-down fuzziness or granularity that you describe?"

"How would I know?" answered her companion, staring at her wide-eyed (though that, of course, was the only way he *could* stare). "Why are you asking me? I'm just a scarecrow. I can only tell you what I observe, and that is how things are."

Slightly abashed, Dorothy looked away at the surrounding electron fog and noticed that there was a sort of flickering running over it, rather like the effect she had sometimes seen in a poorly adjusted television picture or an old movie film. When she commented on this, her acquaintance thoughtfully nodded his straw-filled head.

"That is a sign of the way the *phase* of the amplitude is varying all the time."

"And what is the phase?" Dorothy asked with some trepidation.

"I think we can best answer that by going on down the road to see the *Interference Dancers*. I believe they are about to perform one of their popular Interference Dances, and it is not very far." He set off briskly along the Building Block Road, swinging one leg smartly after the other as Dorothy scurried along beside him. Soon they came to a group of houses that partly surrounded an open square. On the square, a company of flamboyant figures were forming up into lines. As Dorothy arrived, one of them left the group and came up to her.

The new arrival appeared to be a woman in fanciful clothing with flowing skirts. Many scarves and streamers flew out from her costume. At least this was how she appeared, but it was rather difficult to say for sure. All in all, she made a strange figure indeed to Dorothy's eyes. Her hair and all her various scarves floated out from her body in all directions to fade tenuously into the surroundings. Though a striking figure, she was strangely ill-defined, particularly around the edges. When Dorothy looked toward one side of this singular figure, where its form became more and more nebulous, she noted that she seemed still to be looking directly at the apparition, though it now appeared much fainter and less substantial. All in all, the figure was so remarkable that the girl found herself staring more intently than was quite polite. She was startled out of her inspection when the new arrival addressed her.

"That's right, Dearie, I am an *amplitude*, a member of the popular group of 'Interference Dancers.' You may have heard of us and our spectacular probability distributions." She dropped a flowing curtsy with a (probably) self-satisfied expression.

"As you are no doubt beginning to learn, things here are not tediously restricted to being in one definite place and not anywhere else. We spread ourselves about, as you might say. We are distributed. The amplitude for anything has associated with it a probability distribution. This gives the probability of actually *finding* that thing at one position or another. We are *more probably* in some places than in others, but you *might* find us where you least expect. Well, perhaps not where you *least* expect, for there will be negligible probability of our being *too* far out, but we do like to be a little unpredictable."

AMPLITUDES AND PHASES

In classical physics, probabilities combine directly, and adding more options always *increases* the probability.

In quantum physics, an amplitude gives the most basic description, and this has a *phase*. Amplitudes may combine to add *or* subtract, with the result that including another amplitude may either increase *or* decrease the size of the total amplitude and the resultant probability.

Probabilities are given by the amplitude's size only (in fact, by the square of the amplitude) and do not depend on the overall phase.

"If you extend so far out, surely you must collide with one another. Doesn't that interfere with your dancing when you are all trying to work as a group?" asked Dorothy, this being for some reason the first consideration that came to mind.

"Of course it does! That is the whole point of interference dancing. It is the interference patterns that make it so spectacular. We try to make the most dramatic use of our phases, you see."

"That was what I was supposed to ask about!" remembered Dorothy. "What do you mean by your phases? What is a phase?"

"Well, that is a little difficult to explain. Let me try to show you." So saying, she reached within the voluminous folds and layers of her outfit and produced what at first sight Dorothy took to be a tambourine. It was a large round object, but on closer inspection, she saw that it did not have any little bells around the outside. Instead, it had in its center a large arrow or pointer, rather like a compass or the hand of a clock. This arrow was continuously circling, pointing now up, now down, now to one side or the other.

"This is my phase," the amplitude informed her. "It is in a sense a direction, but it does not lie along any particular direction in space. It is a sign of our orientation toward one another. You might think of it as more like the color wheel used by artists to indicate complementary colors."[2]

Dorothy could see that the face of the object was now marked out with a circle of colors, each facing a color that was complementary, or opposite, to it. On one side the colors blended through red, green, and blue. On the other side of the circle, opposite to these, were their complementary colors: cyan, magenta, and yellow. The pointer continued to circle the face, moving steadily from one color to the next. As Dorothy watched, the red and cyan areas detached themselves from the rim and drifted in toward the center of the circle. In the center they overlapped, and the two combined to give a white region. This matched their surroundings so that they both vanished from sight.

"My phase is not constant. It is steadily changing all the time. We amplitudes are all a bit AC/DC. Our current phase determines how we combine with other amplitudes."

"Really, how can it do that?" asked Dorothy with interest.

"When two amplitudes meet, their relative phases settle how they should combine. If both have the same phase, they add. They are then twice what

[2] Please note that this is a description of ordinary colors, used as an illustrative analogy. Ordinary colors are *not* the colors possessed by quarks. Despite their name, the latter "colors" have nothing to do with color as we know it, and they have little to do with phase.

they would be separately and four times as intense. The probability that is associated with any amplitude is given by squaring the *size* of the amplitude, which is to say multiplying it by itself. The probability does not depend at all on the overall phase, although, as I say, individual phases determine the way amplitudes add. In the case where the amplitudes add *in phase*, the *probability* for the combination of two has become *four times* as big as for one amplitude alone. Finding another amplitude of like phase gives one a great boost, I can tell you. We both feel twice as tall," she exclaimed proudly. "You have to experience it to understand. It doesn't matter what our phases are—up, down, or sideways—as long as we are both the same."

Here she reflected a moment, and her tone grew more subdued. "At the other extreme, I may meet an amplitude with opposite phase, one that exactly complements, or opposes, mine. That is a very deflating experience because we cancel each other out. Our sum can be so much smaller than the parts, because what we are subtracts from the other's being, and nothing remains. It is one of those relationships where we cannot live together, and so we destroy each other, at least for as long as we stay together. After we have parted, we return to what we were before, perhaps sadder and wiser—who can say? This last case is known as destructive interference, when two amplitudes may cancel each other completely."

PHASE AND INTERFERENCE

Classically, if there are more ways in which something may happen, then it is *more* likely to happen.

In quantum physics, when two amplitudes of equal size combine, they may, at one extreme, combine *in phase* to give twice the amplitude—*four* times the probability for one amplitude alone. This is *constructive interference*.

At the other extreme, two amplitudes that combine *out of phase* subtract to give *zero* probability. This is *destructive interference*.

Including more amplitudes *may* reduce the overall total at some points.

In general, the relative phases of two amplitudes will vary from place to place, and this tends to give both regions of increased and regions of reduced probability—the typical light and dark bands of optical interference patterns.

The probability over the entire region *averages* out to a value that is just the sum of the probabilities for each amplitude on its own, but the distribution may be quite different.

"That all sounds very strange to me" admitted Dorothy. "I must confess that I still do not understand what this phase may be and why it should have such an effect on your encounters."

"As to *why* it should so affect us, well, who can say why the world is as it is? It still remains the case that when two amplitudes combine, the relationship of their phases determines whether they will cancel each other out, enhance each other, or simply pass with little apparent effect. It is not unlike the situation that you may observe with slight waves or ripples on the surface of a lake. As a ripple travels along, the water rises and falls to give successive peaks and troughs that mark the varying phase of the passing wave. Say such a wave passes on either side of a post rising above the lake surface. Then where the parts of the wave come together again, they may combine *in phase*, to give a greater disturbance in the water, or they may be *out of phase* and cancel. In the second case, the water at that point is still and does not move up and down at all.

"Anyhow, we are about to begin our Interference Dance," she concluded. "Come and see how much it depends on interference for its effect."

She hurried back to her troop of dancers, and they formed two lines across the clearing. The dancers were all exotically dressed, each in flowing garments that faded into insubstantiality, so Dorothy assumed them all to be amplitudes of some sort. Some wore costumes that looked vaguely medieval. Some sported heads like hobbyhorses. Some wore elaborate architectural concoctions, like tall baroque towers. Some were dressed in robes of royal sumptuousness, with crowns upon their heads. In every case, the figures spread wide and faded from sight at their peripheries.

Abruptly each line began to dance toward the other. Dorothy could hear no music, and she might have wondered how they kept in step except that she could see they did not. Each one seemed to be dancing to his or her own rhythm, and the phases of all the dancers were flickering around at different rates.

The lines of dancers passed through each other. This did not mean, as you might normally expect, that the individual dancers in each line passed by their opposite numbers through gaps in the other line. No, in this case, the dancers actually *passed through one another.*

As they passed, each amplitude was *superimposed* on the one opposite. Sometimes the two had the same phase, and at that the line surged upward with an intensity several times its original value. At other points the phases were in opposition, and there the dancers vanished completely. Dorothy was horrified at this, but she then was relieved to see that as the lines passed and separated, the dancers were restored whole as before. Again and again, figures would meet and dissolve, though they always appeared again as soon as

R Gilmore

they separated. As Dorothy watched a horse-headed figure meet with a tall, grotesque tower and saw the two demolish each other, she was reminded of a game of chess, though it was a strange game in which *both* pieces were taken and both were thereafter returned to the board unscathed.

To and fro the lines of dancers swayed, splitting and turning so that they passed through each other again and again. On each pass Dorothy saw them rise to great heights in some places and sink to nothing in others, so that the square was filled with complex and ever-changing patterns of peaks and hollows that rose and fell as they danced together.

When they had finished, the amplitudes all lined up and bowed (probably). Dorothy clapped with enthusiasm. It had been a marvelous sight, and at last she felt that she had *some* appreciation of the effect that phase might have on amplitudes when they interfere with each other.

The dancing group left the square, and at first glance it subsequently appeared deserted. Then Dorothy noticed that there was a small market stall of some sort over in a corner. She asked an interested spectator who had been watching the dance what this was and what, if anything, it sold.

"Why, that is the town market! This is the market square, so naturally the market is here."

"But doesn't a market usually have a lot of stalls, selling all sorts of things?" asked Dorothy in puzzlement.

"Indeed it usually does, and so does this one. There are scores of stalls here, selling practically everything you could possibly want."

"But I only see one!" protested Dorothy.

"Ah no! You actually see them all, but because they are all in the same place, you see a *superposition* of them all. This is in fact the State Fair for the locality, so there is a whole superposition of states. That is quite a common situation hereabouts. There are stalls here for everything imaginable. That must be the case because, after all, there always *will* be an amplitude present for everything that is possible. You can see that from the regulations on the town notice board," pointed out the helpful citizen.

Dorothy had not observed it before, but now that it had been pointed out, she saw a wooden board supported on posts and with various notices spread over its surface.

What is not forbidden is compulsory.	Options are not optional.
You *must* have your cake and eat it.	Everything that can happen will.
Deny yourself nothing.	You *can* have it all.

"Some of that is perhaps going a bit too far," remarked the Scarecrow, who had come to stand beside Dorothy. "'Everything that can happen will' means that there must be present an *amplitude* for anything that can happen. Because amplitudes have their associated probabilities, there is some *probability* that anything that can happen will happen. Put like that, of course, it becomes a truism."

SUPERPOSITION

The principle of superposition says that for anything that *can* happen, there will be an amplitude present.

This is a bit like the classical statement that there will be a probability present for anything that *can* happen. The remarkable thing about the quantum amplitudes is that the various possibilities can *interfere* and affect one another.

"Maybe so," returned the obliging bystander, "but as far as amplitudes are concerned, it remains true that there will be an amplitude present for any possible situation, however unlikely. It might be rather a small amplitude, but it will be there. These amplitudes will actually all be present, and the different possibilities may interfere with one another."

"If anything possible is compulsory," pursued Dorothy, trying another approach, "why aren't there market stalls all over the area? That would be equally allowable, would it not? And a lot more reasonable," she added.

"Oh, but it isn't allowable at all," cried the knowledgeable local. You can't have looked closely at the notice board. Didn't you see the small print at the bottom?"

Dorothy looked again, and indeed there was a notice that she had overlooked, printed in particularly small letters.

TOWN BY-LAW
All market stalls must be located *only* in the designated position.

As though to emphasize this point, there came a clatter as another trader rode up in a wagon. Seeing all the empty space, he parked some distance from the rest and began converting the wagon into his stall. Scarcely had he swung the wagon into position, however, when a short column of men-at-arms came trotting out from between the houses. They surrounded the newly arrived merchant and quickly persuaded him to observe the restrictions laid down. Despite his indignant protests, he was smartly herded into his proper location, where he immediately blended in with all the other stall holders.

Taking their leave of the informative bystander, Dorothy and the Scarecrow made their way over the empty space to the lonely looking market stall two-thirds of the way across, except that, despite appearances, it was apparently not just one stall but a whole market, Dorothy reminded herself.

As soon as they arrived, a merchant accosted them.

"Hello, Sir and Miss. What would you like to buy today? There are plenty of bargains to be had, for trade is not very brisk just now." This last at least seemed to be a fair comment; the girl and her companion were the only two folk she could see anywhere near.

"I think I might like something to eat if, if you have anything like that," she replied cautiously.

"Oh, certainly. How about some lovely pork chops or a nice bit of steak?" The proprietor had become clearer, and Dorothy saw that he had a stolid red face and wore a striped apron and a straw hat. Behind him she could see that the stall was hung with cuts of meat of all descriptions. Obviously, he was a Butcher.

"No, thank you," said Dorothy. "I don't feel like meat at the moment, and I wouldn't have any way to cook it. Is there anything else I could have?"

"Certainly, my dear. How about a nice sesame roll, or perhaps some raisin bread?" Remarkably, the stall holder had changed in response to her inquiry. Now he was fat, with a round, pasty face. He wore a great white apron and a soft white hat, and behind him were piled breads and cakes of all descriptions. He was now a Baker.

"I'm afraid that they look a bit heavy. I should prefer something light."

Once again the stall changed. Now the merchant was tall and thin, with a rough canvas apron coated in grease. Behind him the stall was stacked with candles in holders of all descriptions.

"No, no," Dorothy said to the Candlestick Maker. "I do not mean that sort of light! I was thinking of something quite different."

On the words *quite different,* the stall in front of them began a mad shuttle through various possibilities. There was a stall selling vegetables and flowers, operated by a woman in an appropriately flowery dress. This was promptly replaced by a stall selling dresses, run by a tall, thin lady in a dress that looked positively dowdy by contrast. Next appeared a stall displaying

R Gilmore

books, whose small, balding proprietor was so engrossed in reading one of his own books that he did not notice Dorothy at all.

These came in rapid succession, and all were quite sharp and clear for the brief period during which they were in evidence. They were interspersed with others that were much fainter, representing less probable traders. There was a blacksmith, with a red-hot furnace visible behind him. There was a prim-looking man who peered out at Dorothy from behind a heavily armored glass partition; she was not quite sure what he did. Very faint and obviously very improbable, she fleetingly saw a man in a boldly checked suit. He bore an expression of transparent sincerity and stood in front of a placard that read

Honest Bob's
Midsummer Madness!
Unrepeatable bargains!
We make a *loss* on every automobile
sold!

"Stop, stop!" cried Dorothy. "I don't want any of that. I think I should like some candy."

No sooner said than the merry-go-round of options stabilized to a single booth, in which were displayed all sorts of delectable confectionery: lollipops, peppermints, toffee, chocolates, and candy bars of all descriptions. This stall was attended by a large, avuncular man with a rather sly expression. Dorothy was not sure that she cared much for him, but she found his wares attractive enough.

"Might I have some of that butterscotch?" she asked, pointing to the contents of a glass jar near the front of the array and offering a dollar bill in payment.

"Certainly, my dear! I am afraid, though, that I cannot give you any change. I unwisely set my cash box in the same position as that of another trader, and unfortunately they had opposite phase, so they have interfered destructively and my cash box has vanished. It will reappear when the market breaks up and the amplitudes separate, but for now I cannot give change. Would you like one of these peppermint sticks, perhaps? Some of them have interfered constructively with the Candlestick Maker's candles, so they are enormous."

"No, thank you," replied Dorothy politely but firmly. "I should not care to think what they might taste like. On second thought, I am not so sure that I feel like any candy. Good day to you," she added courteously as she backed away from the market as quickly as she could without appearing too rude.

Just then she noticed that the Building Block Road continued from the far side of the square, and the Scarecrow was quite happy to accompany her as they set off down it.

"Well, that *was* peculiar," remarked Dorothy as they walked along. "I know that I have been told how states will superimpose on one another, but I have been to State Fairs back in Kansas, and there all the stalls are completely separate."

"So they would be back in Kansas," responded the Scarecrow. "There, all the stalls and merchants are large, complex objects that contain huge numbers of atoms, and any interference effects have so averaged out that you see no sign of them. In Kansas you may see interference of electrons and photons, but not of much larger objects; of states, but not of shopkeepers. It is different here in an allegorical environment. Here, anything or anybody you meet may exhibit quantum behavior to emphasize a point."

As they walked on down the road together, Dorothy tried without success to imagine what it must be like to exist solely in order to illustrate some theoretical point.

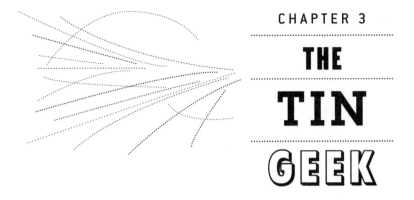

CHAPTER 3

THE
TIN
GEEK

Dorothy and the Scarecrow walked on down the road, which shortly there-after entered a wood. The visibility between the trees was good, and some way off Dorothy could see a glint of sun on bright metal. This intrigued her, for it seemed out of place in such natural surroundings, and she led her companion off the path and up to the spot she had noticed.

She found there, sitting motionless on a fallen trunk, an astonishing figure that appeared to be constructed entirely of bright tinplate. He had thin, shiny limbs with articulated joints, a cylindrical head with a conical top, and a body like a tin box. The only features on his face, if that word may be applied to the front part of an otherwise featureless cylinder, were two large lenses that, for the moment at least, looked dull and lifeless. On the front of his chest there was a screen, rather like a small television screen, on which Dorothy saw the following cryptic message:

>insert system disk in drive A

"What can that mean?" Dorothy asked herself aloud. "What is a system disk, I wonder?"

Because he was the only individual present in any condition to answer, the Scarecrow replied, "Perhaps that is it on the ground beside him." Sure enough, there was something lying there, close to his motionless hand. Dorothy didn't think it looked like any sort of disk because it was square, but it was the only candidate visible. "I wonder where it is meant to go?" she asked.

Again, it was really a rhetorical question, but the Scarecrow, ever observant, answered her. "There is a sort of slot underneath the screen. See if it
will fit there." Dorothy did so and, after a couple of tries, managed to make
it go in. As soon as it was pushed fully home, there was a whirring noise, and
a sequence of unenlightening messages flashed across the screen. A dull
glow sprang up behind the figure's eyes, he began to move his arms, and
then he rose unsteadily to his feet.

"Thank you. Thank you," he said. He spoke in short, unemphatic sen-
tences, and his voice sounded monotonous and mechanical. "I might have
sat there forever. But you came along and reloaded my system. You have certainly saved my life. Or rather you have restored it."

"Not at all," said Dorothy, her own voice shaking a little, uncertain what
to say or indeed what she had done. "How did you come to get into that
condition?"

"It is rather a dismal story. I was a computer enthusiast. I still am, in a
very significant sense. I enjoyed nothing more than working on my computer. Whenever improved hardware came out, I was the first to fit it." His
eyes glowed with enthusiasm as he spoke. Dorothy had often heard this expression used, but never so literally! The dull light behind his lenses had
brightened, and now his eyes shone brilliantly. "As time went on, more and

R. Gilmore

more updates came out. I spent more and more time replacing components. Without really considering whether they needed replacing, I am afraid. I did not even consider what they were or what they did. Then I made my terrible discovery. But it was too late."

"What discovery was that?" asked Dorothy, trying to sound appropriately solicitous rather than simply intrigued.

"I found that, in my enthusiasm, I had upgraded myself. I am now essentially a computer. A very up-to-date one. But I am still a computer. I was sitting out here when there was a thunderstorm. The current surge from the lightning wiped out all my programs. So there I was, or rather wasn't, until you reloaded me. Thank you again." The flat, reverberating voice stopped abruptly.

"Think nothing of it," replied Dorothy. "I am sure you would do the same for me," she heard herself saying before she realized how ridiculous that was. "How long have you been here?" she asked to cover her confusion.

"I have been here for three months, two weeks, five hours, six minutes, and forty-seven seconds, according to my internal clock. Except, you see, I have not been here. I had ceased to function. I have no memory." But after a moment he seemed to reconsider. "Actually, I do have a memory. It is non-volatile and very extensive, if I may say so. I suppose it must be quite out of date by now." Here Dorothy thought she detected a trace of sadness, though it did not show in his voice. "I know how long it has been since I shut down, but I do not know anything that has happened in that time."

"Why don't you come with us?" offered the kind-hearted girl. "We are going to see the Wizard of Quarks, and I am sure he can tell you everything you need to know. I understand he is very wise and well informed."

The Tin Geek agreed enthusiastically (though again, you could not tell this from his voice), and they all set off along the Building Block Road.

The path, which had never been totally straight, now began to twist from side to side and also up and down. As they were plodding up a fairly steep hill with a sharp bend in the road at its crown, there was a sudden clatter of hooves and a rumble of wheels on the uneven surface, and a furiously driven wagon came leaping and careening over the crest of the hill. It swerved frantically as the driver saw the sudden turn on the road—but too late! The wagon veered wildly off the track and slammed sideways into a substantial tree. The horse, after pausing to shake his head and utter a loud neigh of disgust, broke free of his traces and galloped off through the wood, followed frantically by his erstwhile driver.

Dorothy and her companions were not able to pay much attention to the subsequent fate of either horse or driver, because the wagon had burst asunder as it collided with the tree, and its various contents were tumbling down the hill toward the spot where they stood.

They watched in some trepidation as dark objects came bowling and leaping past them. As far as Dorothy could tell, whatever else there might have been in the reckless driver's load, he had been carrying a great number of large lumps of coal. She watched these rushing past her and thought that it must have been a *very* large amount of coal indeed, for the flow of rapidly moving pieces continued for a remarkably long time. Many of the lumps had come to rest in the road nearby, and now and then a new arrival would collide with a stationary one. She noted with passing interest that what happened after each such impact depended on the relative sizes of the pieces of coal involved.

On one occasion a large mass struck a small one, and both continued forward, the smaller now moving more rapidly than the large. At other times a small mass struck a larger target. The larger would be propelled slowly forward, while the smaller newcomer rebounded backward. Occasionally two masses of equal size would collide, and this resulted in a total exchange of roles, the stationary one leaping forward at the speed with which it had been struck, the other coming to a complete halt in its place.

"There you see the conservation of momentum," the Tin Geek remarked from close behind Dorothy's ear as she watched the various collisions. "Have you ever asked the question 'What is the true measure of motion?' "

"I cannot say that I have, but it is obvious isn't it?" replied Dorothy. "It is movement itself, the rate at which something moves. It is what you would call speed, I suppose."

"No, it is not really. Momentum is more significant. Momentum is the speed—the velocity—multiplied by the mass of the thing concerned. If you look carefully, you will see. Sometimes an article will bounce back after it strikes a tree. It may then collide head-on with an oncoming object. The two share their motion. How it is shared depends on their masses as well as on their velocities."

Dorothy looked as she was bidden and saw that, indeed, there were instances where an item hurtling along collided with a tree and rebounded into the path of another. When two objects of comparable size met, they both moved on in the direction of the faster. In another case, where a heavy mass struck a smaller mass that was moving only a little more quickly, they both moved, after the collision, in the direction of the larger one, but more slowly. In one case a smaller and faster mass collided with one slightly larger and slower, and both came to complete rest.

"You see, momentum is conserved. The product of mass and velocity multiplied together is preserved after a collision."

"How can you say it is preserved?" protested Dorothy. "In that last case I saw, there was motion before, but afterward there was none. Both had stopped. How can any measure of motion have remained the same in such a case?"

"Momentum is a vector quantity. It has a direction as well as a magnitude. When two momenta with opposite directions are added, they tend to cancel. If they are equal and opposite, they cancel exactly. The total momentum will be zero. This is true however fast each component was traveling. The total momentum is the same as when both are at rest."

"But," protested Dorothy, sure that something was wrong here, "something must be different. I cannot believe that two things moving very quickly have just the same motion as when they are at rest. That does not make sense. It cannot be true."

"It is not true. The total momentum is the same. But momentum is not the only measure of motion. The total kinetic energy will change. Kinetic energy has no direction. It is always positive and it adds. The two masses had kinetic energy before they collided. After the collision they had none."

"So energy is not conserved when things collide. Is that it?" Dorothy was trying hard to get this straight.

"No, that is not it. Certainly not. Energy is conserved. But kinetic energy is not. Energy can convert from one form to another. Kinetic energy is the energy of a moving object. There are other forms of energy."

At that moment there was a sudden explosion from the shattered wagon. Something, presumably whatever had been burning all that coal, had chosen that moment to blow up. There was a whistling noise, and a large iron pot came sailing through the air. It landed with a loud crash in the middle of the road close to where they were standing. The Tin Geek walked over to it and, with a visible effort despite his metallic strength, lifted it. There was a large dent in the road surface where it had struck.

"That was a near miss." He remarked. "It was also an illustration of the conversion of energy. The explosion released chemical energy in the coal dust. This converted to kinetic energy, the energy a moving object has simply because it is in motion. That energy hurled the pot into the air. As the pot rose, it slowed. The kinetic energy it had was converted to potential energy of gravity. As it then fell, it gained speed again. The potential energy converted back to kinetic energy. On impact, some of the energy was converted to sound waves. The rest went into distorting the road. The total energy was the same throughout."

"The pot has made rather a mess of the road, though," said Dorothy thoughtfully. "That is a shame."

"No, it is quite normal. Most roads have a few potholes in them."

"Certainly, it is quite normal," agreed the Scarecrow, who had been watching throughout. "It has all been normal. Much too normal in a way. You have been involved in the 'illusion of normality' again, and you have been seeing moving objects as you would see them if you were back home in your large-scale world."

Dorothy was finding this discussion so interesting that for the first time, mention of home didn't make her feel sad. "How do things differ here when they move? Surely movement is movement, whatever the scale of things."

"Perhaps movement is movement, as you say, but on the scale where amplitudes and quantum effects are obvious, it all *appears* rather different. Look again at the passing objects and now see their amplitudes."

Dorothy looked at the stream of hurtling masses (they did seem to be coming from the wrecked wagon for an unconscionably long time!). Now, instead of perceiving them as individual and distinct, she saw each as a long blur that extended past her down the road. Along the length of every one was a varying pattern that she recognized as being like the variation of phase that she had last seen flitting across the capering amplitudes at the Interference Dance. She saw bands of varying phase that ran the length of each amplitude. For some the bands were widely spaced; for others they were much closer together.

"Each one has a different momentum. That means it has a different wavelength. The wavelength is the distance between peaks. The peaks are the points where the phase is at the top of its cycle. The wavelength is greater for some than for others, because of differences in their momentum. The faster ones have smaller wavelengths. The peaks in the phase are closer together," explained the Tin Geek, speaking at some length for him.

"That does not seem right," responded Dorothy as she tried to visualize this. "Surely the faster the motion, the *farther* the peaks should be separated. Shouldn't they be more spread out if they are moving more quickly?"

"They are spread out. The separation has increased with the speed of the particles, and their amplitudes are more spread out, but that is not the dominant effect. As the particles go faster, their energy increases, and this means that the frequency of the amplitude has increased also. The frequency is the number of times the phase cycles per second. This increase is in proportion to the energy, and the kinetic energy increases more quickly than the speed or velocity of the object. The kinetic energy increases as the square of the velocity—that is, as velocity multiplied by itself. If the speed were to double, the kinetic energy would increase by a factor of four. That is an even greater change, and the increase in frequency dominates. So you see, the phase changes more quickly and the bands of phase are in fact closer than before."

"So this phase that you tell me about is changing all the time?" inquired Dorothy, just to check that she had grasped it.

"That is correct. And so the wavelength decreases as the momentum increases. The wavelength is the distance along a wave between successive peaks in its phase. At the same time, the frequency increases with the energy, as I have just said. The phase cycles around more times in a second. As you see, amplitudes do not simply sit still."

DE BROGLIE'S AND EINSTEIN'S RELATIONS

These relations are remarkable basic facts of quantum physics. The relations are observed to be true, although there is no way in which you could have predicted that they should be.

They relate the wave-like properties of particle amplitudes, such as wavelength and frequency, to the so called kinematic properties of the particles, such as momentum and energy.

De Broglie's Relation: The wavelength, λ, is inversely proportional to the momentum, p.

$$\lambda = \frac{2\pi \, \hbar}{p}$$

Einstein's Equation: The frequency, ν, is proportional to the kinetic energy, E.

$$\nu = \frac{E}{2\pi \, \hbar}$$

Why is this so? Who knows?

Dorothy looked where the Tin Geek was pointing, at one of the lumps of coal that had come to rest and was lying by the roadside. She was not too surprised to see that it was indistinct and became increasingly fuzzy toward its edges. She had come to expect that. What she had not expected was that the object seemed to boil with shifting phases. As she watched, it gradually spread out farther and farther, growing more extensive and diffuse as each moment passed.

"What is happening?" she asked. "I know you have told me how amplitudes are always fuzzy and spread out, as that appears to be, but why does it change so?"

"Because it is not a single wave of definite wavelength. It is a wave packet. It is a packet of waves of different momenta. No object can be completely at rest. Some of the waves in the packet must have momentum. This means that they will move and that all will move differently. Because of this, the amplitude keeps changing."

"But why?" protested Dorothy, feeling that every question she asked was answered with some completely indecipherable new statement. "Why should an amplitude have all this motion and not be completely at rest? That seems quite unreasonable to me."

"Be that as it may," interjected the Scarecrow, "it does follow from what you have already been told. I'm sure the Tin Geek is eager to explain it to you."

As far as Dorothy could see, he did not look particularly eager. But then, she reminded herself, it must be very hard to look eager when you have a featureless tin cylinder instead of a face.

"A particle that has a definite momentum has a definite wavelength," the Tin Geek began. Dorothy could not see that this was particularly relevant, but at least it was something she had already heard. She waited to hear what else her tin tutor had to say.

"Its amplitude is a featureless wave, like this." As he spoke, the screen on his chest lit up to show a wavy line that ran all the way across it. "This indicates how the phase varies along an amplitude. But it is not, in fact, a very good indication. The phase actually circles round and round, not just up and down."

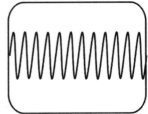

"You remember what the amplitude told you before the Interference Dance?" whispered the Scarecrow. "Phases always cycle round and round like that."

"The variation of phase is better illustrated by that cycling phase tree over there," continued the Tin Geek with mechanical determination. He stepped away from the path and indicated a rather peculiar growth in front of a vertical rock face. The tree, if that is what it was, had a straight central trunk and bare branches that stuck out at right angles. It was these branches that were so peculiar. All were exactly the same length, and they wound up the trunk in a rising spiral. At the lowest level they jutted out to the left; a little farther up they had swung around to point directly toward the watchers. Farther up the spiral the branches protruded to the right, and after another half-turn they pointed to the left again. This spiral of branches continued up the tree until it was lost from sight in the foliage of other trees. The low sunlight cast a shadow of the phase tree on the rock face behind, and in this shadow, the branches that pointed toward or away from the rock were foreshortened. The outline of the shadow branches gave a wavy line, much like the one the Geek displayed on his chest screen.

"The proper form is not easy to show on my screen because it is flat. I have shown merely a projection, a sort of shadow or side-on view." The Tin Geek's matter-of-fact tone held no trace of apology. "The probability of finding the particle at any point does not depend on the phase. Only the size of the amplitude matters. For this wave, the amplitudes (like the branches of the tree), are all the same length. The probability is the same everywhere for such an amplitude of definite momentum and wavelength. The particle is equally likely to be found anywhere along this wave. Its amplitude extends far on either side. The particle could be found anywhere along its infinite length, and so it is nowhere. There is no point where you could say, 'The particle is most likely to be here'."

"But a particle must surely be somewhere!" objected Dorothy. "Perhaps, as you all tell me, its position is indeed uncertain. It may be a little fuzzy, but it must be somewhere!"

Here the Scarecrow joined the discussion. "I am something of an expert on observation, and I can tell you that when you do have a particle with a definite momentum, its wavelength is constant and there is no way of telling from the amplitude where the particle is. And because the amplitude is all you can know, that means the particle could be anywhere."

"That's silly!" Dorothy insisted. "The position of a particle may be a bit fuzzy, but you can still have some idea where it is. If I have my lunch in a box, I may not know exactly where all its atoms are, but I do know that they are in the box and not spread all over Kansas."

"Oh yes," replied the Scarecrow agreeably. "That is perfectly true. But what I said is the case for a particle of a single, unique momentum. You could say nothing about the position of such a particle, but the atoms in your lunch—and indeed other particles in general—do not have a single, unique momentum. They have a mix of momenta, just like the dancers in the Interference Dance that you watched. And in just the same way, their amplitudes vary from place to place because of interference."

"That is what I am saying," the Tin Geek's monotone resumed. "The amplitude is not a single momentum wave. It is a wave packet. Until you measure the position, you can speak only of probabilities. You might say a particle is 'here' rather than 'there' if its probability is greatest 'here.' There must be a peak in the probability distribution. It must be most probable that the particle is near one particular place. This momentum wave that we spoke of is the same everywhere, and it must somehow be concentrated in one place. It must be large there and must fade away to either side. This may happen because of interference. See the effect of adding two waves of slightly different wavelength."

As he spoke, a second wavy line appeared below the first on his chest display. It looked very much the same, except that close examination showed its peaks were just a little farther apart. This was made clearer as the second wave moved up his screen until it overlapped the original wave, and then he showed the sum of the two combined. Now the size of the up-and-down wiggles changed from place to place. In the center, where the two waves had been in phase, they added, and the result was much larger than before. The wiggles shrank in size away from the center until, some distance from each side, they faded to nothing. This was where the two 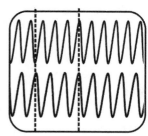 waves had become so out of step that they were completely out of phase and had canceled each other. Farther out on either side, the total wave began to grow again until it was once more as large as at the center. This was where the two waves were out of step by a complete cycle and so interfered constructively once more.

"Now the probability is greater at the center. Unfortunately, however, it is just as great at a series of positions on either side of the center. With such an amplitude the particle might not be everywhere, but it is equally likely to be at a succession of different places. An infinite succession. These series of peaks that you get from adding two waves are known as 'beats.' A particle that is to be in one  particular region must have only one peak in its amplitude. Somehow it must lose the other peaks. This can happen if more waves are added."

On the display, more waves appeared and rose up to combine with the original sum. Each was in phase at the center, and there the peak grew larger and larger with each new addition. The new waves all had different wavelengths, so on either side they got out of step in different places and reduced the other peaks. Faster and faster, more and more waves were added on the display, until only the central peak remained. The oscillations on either side now shrank smoothly in size, and at positions far enough removed from the peak, they declined to virtually nothing.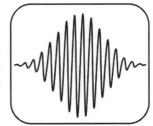

"An amplitude like this one tells you where a particle is. It is most likely to be in the peak region. Only there is the probability large. If the peak were narrower, you would know even better where the particle was. You can have such a peak only when waves of different momenta have been combined. To get a narrower peak, you must have an even wider range of momenta. Then they get out of step more quickly."

HEISENBERG'S UNCERTAINTY PRINCIPLE[1]

This is another of the more remarkable quantum results.

The de Broglie relation between momentum and wavelength is an experimentally observed fact. You could not have argued in advance that it *should* be so,

---

[1] I personally feel that this should be called the Uncertainty Relation. It is not really a separate principle in its own right, because it can be derived entirely from the wave-like properties of quantum amplitudes. There are, however, two schools of thought about this: *I* do not think it should be called a principle, and *everyone else* thinks it should. Accordingly, I refer to it as a principle in the text.

but apparently it is. It has the consequence discussed here by the Tin Geek—that position and momentum for a state are together uncertain. This does not simply mean that *we* do not know what they are. Nature does not know either; the values themselves are actually fuzzy and indefinite.

The Heisenberg Uncertainty Principle connects the fuzziness in position, Δx, and the fuzziness in momentum, Δp.

$$\Delta x \, \Delta p = \frac{\hbar}{2}$$

(It is often written that $\Delta x \, \Delta p > \dfrac{\hbar}{2}$. The "is greater than" sign simply means that it is always possible to be *more* uncertain about something than is theoretically necessary.)

"I don't think I quite follow what you are saying, or rather *why* you are telling me this. What does it all have to do with the way I saw that particle's amplitude behaving? The particle ought surely to have been at rest, but it appeared to spread out in all directions as I watched it."

"That is exactly what our metallic friend has been telling you," the Scarecrow declared "He has just said that no particle can have one single momentum, which means it cannot have a unique speed. That applies to a stationary particle also. You cannot have a completely stationary particle—not, at any rate, if you know approximately *where* it is. Any sort of restriction in position brings with it a spread in momentum. Obviously, if there is a range of momenta, then not every momentum can be zero. The particle will, as you might say, move 'furiously in all directions.' It will disperse."

When the Scarecrow had finished speaking, there seemed little further to be said about the behavior of moving particles, and the three companions walked on down the road. Dorothy was turning over in her mind what she had heard. It all seemed very strange, and she was not sure she really appreciated the connection between position and movement.

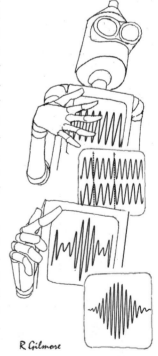

R Gilmore

****

As they journeyed along, they came to a place where a deep ravine opened up on either side of the road. It separated two wide plains that differed greatly in appearance. On one side of the fissure the ground was placid, flat, and featureless apart from a sort of net-like mesh that covered it, rather like the grid on a sheet of graph paper. The other side of the divide was all activity. It was as though a strong wind were blowing across a field of grass. Lines of varying lengths swept across the surface, swinging and swirling in all directions with an intense appearance of motion.

Dorothy commented on the contrast between the two areas, and the Scarecrow, ever obliging, replied. "Those are the regions of *position* and of *activity*, of *being* and of *becoming*. They are close neighbors here, opposing and complementing one another at all times."

Along the middle of the deep divide that separated the two fields there stretched a tightrope, and in the distance Dorothy saw a tiny figure walking along it. As he came closer, she saw that he was a thin, angular person dressed in colorful tights and a leotard ornamented with the ubiquitous symbol $\hbar$. A vertical division ran down the center of the figure, with contrasting colors on either side. His tights and leotard were also differently colored on either side, his hair was dark on one side of the line and light on the other, and his mask-like face was similarly partitioned. Treading carefully, he came toward her, swaying slightly from side to side as he adjusted his balance. He carried a long pole that he held horizontally, and as he leaned one way or the other, it veered toward the side of *position* or of *action*. When the pole was level, it protruded equally on both sides, but when the staff tended to one side, it would shrink on that side, becoming more localized and better defined. At such times, it became much longer and fuzzier on the other side, in compensation.

This individual reached the point at which his tightrope crossed the Building Block Road and stepped onto the path in front of Dorothy and her companions.

"Good morning, afternoon, or evening, as the case may be" he said, with a deep bow that twisted his body into an implausible angle. His voice was an amazing mixture of deep bass and high tenor, and the relative strengths of the two components changed from time to time.

"Don't you know which time it is?" asked Dorothy in some surprise.

"Well, I must admit to being a little uncertain. A degree of uncertainty is the general rule here, as you may know. The position and momentum of a particle will always have, let us say, a certain uncertainty. You have to adjust your thinking to accept this. It must seem strange to you at first, but I specialize in adjusting between latitude in position and latitude in momentum. I am known as the Adjusting Acrobat."

"I have just been told how particles have this spread in position and momentum, but I must say that it still seems very peculiar to me," said Dorothy. "Surely a particle must be located at some particular position and not at any other. It stands to reason."

"If you do not mind my saying so, that seems a very old-fashioned view in one so young" replied the Acrobat. "Positively Victorian, in fact. The idea that everyone and everything should have their proper position and should always be found in it is quite outmoded now. We are living in much more free and easy times. A person's or a particle's position is what he or she makes of it."

"That is not the same thing at all," said Dorothy. "Surely you are talking about *social position*. We certainly do not worry about that in Kansas. The position of something on the ground is quite different. That must be at one point and nowhere else." Dorothy had to restrain herself from stamping her foot to underscore her certainty on the matter.

"Well now, I do not see how you can say that, even in Kansas. Look at that puddle, for example." The Acrobat pointed to an ordinary puddle by the side of the path. It was quite a large puddle, and Dorothy had not noticed it before. She suspected that it had suddenly appeared to join the conversation, but nevertheless it did seem rather ordinary. Dorothy said so.

R Gilmore

"So it is. A normal puddle in fact. Can you show me exactly where the puddle is? Can you point to that single, unique point at which the puddle is located?"

"Well, no, I cannot," Dorothy admitted. Puddles are not like that. They spread over an area, it's true. But it is not the same for particles," she added rapidly, because she suspected she was being tricked into making some statement that the other would then distort to prove his point.

THE SIZE OF $\hbar$

You may feel the Uncertainty Principle is nonsense because you can see for yourself that things do *not* behave like that. A small speck of dust *will* just sit where you leave it, apparently indefinitely, without any spreading such as that described here. The difficulty we all have is in realizing just *how small* $\hbar$ really is.

Consider a speck of dust one-hundredth of a millimeter across, which is certainly as small as you can see, and take the *uncertainty* Δx in its position to be this size also. When you calculate the velocity corresponding to the momentum Δp given by the Uncertainty Principle

$$\Delta x \, \Delta p = \frac{\hbar}{2}$$

you find that in order to observe the speck to move as small a distance as its own diameter (certainly the smallest distance you might conceivably observe), you would have to have watched it since the middle of the age of dinosaurs. Even if you were not distracted in the interim, you would have to allow for other, larger effects, like continental drift.

Such velocities hardly seem worth worrying about, but the effect gets more significant as the objects get smaller. Atoms are *very small*. The electrons in an atom are given velocities about 1 percent of the velocity of light, solely on the basis of the fact that they are contained within the small dimensions of the atom.

"Can you be so sure?" the Acrobat challenged her. "You know about puddles, but how can you be so sure about particles? Their positions may be quite exact on the large scale of your normal experience, but here on the scale of $\hbar$ they are not. They are spread out, as was your puddle. They are uncertain. They are fuzzy. Furthermore, their momentum is also fuzzy and uncertain in inverse proportion to their size. That is the way of particles. It is just how things are here. It is a bit like this."

Whereupon the Acrobat produced a limp object from a concealed pocket and proceeded to blow it up. It appeared to be a large rubber balloon, which he then began to squeeze between his hands. He squeezed it so that it was narrow, but it oozed out tall between his hands. He squeezed it down low, but it spread out wide on either side.

"You see that if I try to make the balloon smaller in one direction, it grows correspondingly larger in another direction. So it is with the position and momentum of particles. If you restrict them to a tight position, then their range of momentum spreads out wide. And if you squeeze the momentum down so that it is well defined, then the position of the particle becomes wide and fuzzy. You always have a tradeoff between position and momentum."

Suddenly the Acrobat leapt into motion. "But enough of this!" he exclaimed abruptly. "I cannot remain localized on this track any longer. It has pushed me too far into action, and I must be gone." So saying, he sprang onto the tightrope on the other side of the path and teetered swiftly away, swaying one moment toward the side of action, the next toward a better-defined position. Dorothy watched the fluctuating, divided figure recede until he was quite out of sight, and then the threesome continued along the road.

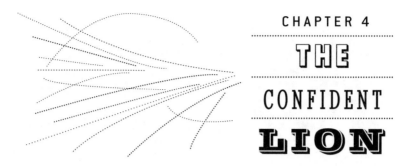

CHAPTER 4

....

THE

....

CONFIDENT

....

LION

Dorothy and her friends walked farther along the Building Block Road. The two complementary plains of *position* and *activity* were soon left behind, and once more their path passed through a wood. This wood was more densely grown than the last. The trees crowded thick together, and the track between them was dark and vaguely menacing. Dorothy felt an irresistible urge to chatter, as though this might somehow counter the forbidding atmosphere. "I am sure it will be interesting when we reach the Wizard and hear all about quarks and things. I have always wondered about them." This was in fact a lie. She had never heard of them before, but it seemed necessary to say something to break the threatening silence.

"I should like to hear more about leptons," stated the Tin Geek. "And bosons," cut in the Scarecrow, just to show that he was still there. "Yes, certainly," agreed Dorothy. "Bosons and leptons and quarks, certainly," she repeated. Somehow after that they all found themselves singing

> *Bosons and leptons and quarks, oh my!*

With linked arms, they marched along the road singing together.

> *Bosons and leptons and quarks, oh my!*
> *Bosons and leptons and quarks, oh my!*

R Gilmore

They moved faster and faster, although the sound of their song served somewhat to dispel the gloom around them.

Suddenly their singing was interrupted.

"I presume you realize that that grouping is inconsistent, in that leptons and quarks are both types of fermions?" a voice boomed out from nearby. They looked around startled and saw, in deep shadow beneath the trees, an even deeper silhouette: a great, dark outline that loomed above them, its darkness broken only by an expanse of shining teeth. They clutched one another tightly as the shape moved toward them. Down onto the path in front of them stalked an enormous lion. As they cowered close together, he stalked right up to them and sat beside them, rearing up on his hind legs. He draped heavy forearms amiably across their shoulders, and gave a wide toothy grin. The light flashed off his horn-rimmed spectacles. In their anxiety, they had not noticed these before.

R Gilmore

"Allow me to introduce myself," he said heartily. "I am, as you may have noticed, a lion. Isa C. Lion, at your service. The C stands for Confident," he added confidentially.

"So your full name is Isa Confident Lion," remarked the Scarecrow.

"Exactly!" the lion roared.

"What was that you were saying about bosons and fermions?" asked Dorothy, talking to overcome her nervousness. "I do not think I have heard anything about them yet."

"Oh dear, that will never do," responded the Lion. "The distinction

R Gilmore

between bosons and fermions is as basic as you can get with particles. I could of course explain it to you myself. You will find there is very little that I cannot explain," he added with no false attempt at modesty. "Just now, though, it might be better for you to hear it from the Falsbadour. He is about to begin his usual performance. Walk this way."

So saying, Isa C. Lion set off on four legs with a smooth, prowling motion. "I do not think I *could* walk that way!" Dorothy said to herself. When she looked at her other two companions, the Tin Geek jerking along like a machine (which of course he was) and the Scarecrow swinging wildly as though on invisible strings, she decided she couldn't walk that way either, so she just followed as best she could.

FERMIONS AND BOSONS

The principal division of particles is between fermions and bosons.

Fermions obey the Pauli Exclusion Principle (which is discussed in this chapter); bosons do not.

Bosons have a spin (a form of internal rotation discussed in the next chapter) that is zero or a multiple of the basic unit $\hbar$, whereas fermions typically have a spin of $\frac{1}{2}\hbar$.

Fermions are *conserved*; a fermion cannot be created without an appropriate antifermion. Bosons are not conserved. If there is enough energy to provide the rest mass (see Chapter 6), a boson may be created on its own.

After they had gone but a short way, they came to a clearing in the woods. Several folk sat around the edge with the expectant expressions of people who anticipate being entertained with no great cost or effort on their part. In the center of the clearing stood a tall, thin figure clad in a bright, medieval-looking costume, with a padded vest and multicolored tights. He was purposefully holding a stringed instrument, most likely a lute, as though he were just about to perform.

"Behold the Falsbadour!" announced the Lion.

"I am sorry," confessed Dorothy, "but I do not know what a Falsbadour is."

"Why, you have heard of a Troubadour, have you not?"

"Oh, yes," said Dorothy, who was rather fond of historical novels. "He is a wandering minstrel who goes around singing ballads about great tidings. I believe that was how news got about before we had newspapers."

"True enough!" replied the Lion. "Now if that is a Troubadour, you can guess what the Falsbadour is!"

"Oh, do you mean that the news of which he sings is *not* true?" asked Dorothy in a whisper, so as not to hurt the Falsbadour's feelings.

"Well, no, it is not that. His basic facts are OK, but he recounts them very badly. Often he is quite unintelligible," the Lion whispered back. Unfortunately, he had not quite grasped the idea that a whisper is not meant to be heard, and his voice boomed out almost as loudly as before. "What is more, he can't sing," he added. The Falsbadour looked distinctly hurt, but he nonetheless pulled himself together and announced that he was about to perform "The Ballad of the Bosons and the Fermions." He began singing in a high, cracked voice:

> *I'll sing you a tale as never does fail in the singing of it to spur me on.*
> *It may seem absurd but wait 'til you've heard of the boson and of the fermion.*
> *The whole Universe rests as you'd never have guessed, not if you tried for a week,*
> *Very much on the way that these particles play their form of the game "hide and seek."*
> *This game's effect is such that a gap you get*
> *As wide as that the poet told twixt Montague and Capulet.*

"Do you understand now what I said about his songs?" whispered the Lion.

"Yes, I think I can see what you mean," Dorothy whispered back. The Falsbadour glared in the direction of the Lion, but went on singing.

> Particles teem of each name, every one is the same (and it
>   interrupts me when you roar so).
> They're as like, it is odd, as two peas in a pod; in fact they're
>   considerably more so.
> So it's true for a start, you can't tell them apart, and if one
>   should change place with its chum
> There is no way to know, the act's hidden and so—that amplitude
>   adds to the sum.
> Any particle or its brother can exchange with any other—every
>   option is entitled to its place.
> Each permutation is included and as I've just alluded, the
>   amplitude will grow and grow apace.

"I am afraid I cannot understand this at all," Dorothy whispered to the Lion.

"I'm not surprised," the Lion roared back at her. He might possibly have *thought* he was whispering, but nevertheless he completely drowned out the weak singing voice of the Falsbadour, who stopped abruptly and glared at him again. All the members of the singer's audience swung around, however, and cheerfully prepared to listen to the Lion instead. "You may know," he resumed, "that all particles of a given type are identical. By this I do not mean just similar, or even very difficult to tell apart. I mean truly identical. An electron is a typical example of a fermion, and one electron is *exactly* like any other. All electrons being absolutely identical, we usually speak of *the* electron, because a description of one completely describes them all, even though there are very many electrons."

"If the electrons are completely identical, does that mean that they must all be doing the same thing? That they must all behave in exactly the same way?" asked Dorothy, who was wondering how things that were exactly the same might in any way differ.

"Oh no, not at all! Every electron is the same in and of itself. They are all identical, and there is no way you can tell them apart, but that does not mean they must behave in the same way. It is a bit like the case of identical twins. One twin may be sitting obediently in class and the other playing hooky at a football game. They are doing different things, even though you cannot tell *which* twin is in which place. Electrons are much more similar than twins. They are completely identical, and you cannot conceivably tell them apart. Even so, they do not do identical things. You have to describe the collection of electrons as a whole by a many-electron wave function, or amplitude. 'No electron is an island,' you might say. The identity of the electrons has important consequences for their amplitude, as you will hear."

The Confident Lion paused to collect his thoughts and then continued. "Electrons are fermions, but the same is true of bosons. A typical boson is the photon, the particle of light, and one photon is the same as any other, apart from its energy."

"Does light have particles, then?" asked Dorothy in some surprise.

"Oh yes, certainly it does. The photon is the particle of light, as I said. I should say, rather, that it is the particle of electromagnetic radiation. This radiation comes in a huge range of energies from very low-energy radio waves to high-energy X rays and beyond that to gammas. Your eyes can see only photons that are within a small range of energies, and a beam of light is a stream of such photons. The photon is as good a particle as any other, except that it has no mass. You might say it is *very* light." As he said this, the Lion roared with laughter. Dorothy had heard of people roaring with laughter but had never experienced it before. She clapped her hands over her ears to protect them until the sound stopped.

"As I was saying," the Lion went on after he had caught his breath, "there is no way, no way at all, that you can tell two electrons apart. The perfect place for an electron to hide is among a lot of other electrons. You cannot tell which was your original electron. Indeed, the very idea of *the original* electron does not mean much, because if two electrons changed places, you could never know. It is always quite possible that two might actually have changed places, and if it is *possible*, then it is *compulsory*. As you should know by now, the overall amplitude must include all possibilities and so must include *both* of these options. The amplitude for a number of electrons must include a term for *any* and *every* pair of electrons to change places with one another, simply because each such change is *possible*."

"That sounds like an awful lot of terms!" said Dorothy in awe.

"Oh, it is, it is. If you have a large number of electrons, as you usually do, then the number of possible pairs is truly enormous, but nonetheless every pair must be accounted for. The overall amplitude must be a sum of terms that includes *every* possible option. That is how Nature works."

The Lion paused for a moment with an expression of vast satisfaction. He had every intention of continuing, but the Falsbadour grasped the opportunity to take up his song again.

When particles are rearranged, what you see is quite unchanged;
the probabilities as such are not affected.
But a question you might raise: "What happens to the phase?"
The answer you may find quite unexpected.

When particles swap, the phase may change. Swap again, and it
changes more.
But swapping twice is no change at all; the initial state it does
restore.
From this you can see with no help from me that one choice for the
phase change is none.
But if it does change, then for double the range, one full cycle is
what must be done.

Bosons arrange that when they exchange their amplitude isn't
affected.
The fermions, though, a difference show, as perhaps you might
now have expected.
They change their phase but in limited ways, the amplitude
changing its sign as
A half-turn bestows an inversion that goes from a plus all the way
to a minus.

You might ask of my patter, does this really matter? The answer—
decidedly so!
For on this change of sign depends your life and mine, in fact the
entire cosmic show.
On it atoms depend or else they'd descend into chaos and utter
confusion,
Which fate was prevented when Pauli invented his Principle of
Exclusion.

Dorothy had been listening as intently as she could. She felt that something important was being said, but try as she might, she could not decide what it was. She said as much to the Lion, who shook his massive head as he looked pityingly at the Falsbadour.

"That is only as I would expect. How that character manages to make a perfectly straightforward argument seem so contorted I do not know. I must offer to lend him my book on the subject." The Lion reached somewhere inside his copious mane and produced a small book that he waved about him

for emphasis. Dorothy could see that it was titled *Fermions and Bosons—The Great Divide*, by Isa C. Lion.

The Falsbadour, once again drowned out by the Lion's booming voice, sat down dejectedly on the forest floor and began to pluck idly at the strings of his lute as he glowered at the Lion. His erstwhile audience again transferred its fickle attention to that confident beast.

"It is all quite simple really," continued the Lion, quite oblivious to the dark looks the Falsbadour was shooting his way. "As I was saying, when you exchange two electrons in a group, it makes no difference that you can possibly, conceivably detect. That means that all the *probability distributions* are unchanged, because they indicate what you may observe or measure. The only thing that might change is the overall phase, because the overall phase does not affect the probabilities. You can find out all about phase in my little book here."

The Lion reached, apparently at random, into the surrounding undergrowth and produced a book that he displayed to Dorothy. It was titled *Do Amplitudes Ever Get Phased?* by Isa C. Lion. "Perhaps you would like to glance through it? No? Well, maybe later. It can be a bit difficult to find the right pages when you want them."

The Lion's enthusiasm was undiminished by Dorothy's reluctance to read his book. "Anyhow, in the present case the number of ways in which the phase can change when two electrons switch over is pretty restricted. Whatever may happen when you exchange two electrons, you know that when you exchange them *twice*, you are right back where you started. In that case there has been *no change at all*, and everything, even the phase, must be as before. That leaves you with two options. One is that the phase does not change at all when two electrons swap places. Clearly, in that case *everything* will be just the same as it was before the electrons were exchanged, and everything is equally unchanged after two such exchanges. The other possibility is that the phase has rotated around by a complete circle after the two successive exchanges have taken place. The phase would then be pointing in the same direction as it was before, and everything would be unchanged. The amplitude would be just the same as in the former case, but remember that I am now speaking of what happens when a particle is exchanged twice. After *one* exchange, the phase would have rotated by only half a turn and would be opposite to its original state. Amplitudes in this case would be reversed, being in effect multiplied by minus one.

"So you see," continued the Lion, sitting down comfortably and grinning at the members of the Falsbadour's former audience. "You have two options, and these options correspond to the two classes of particles: bosons and fermions. The amplitude for bosons does not change in any way when they change places, whereas that for fermions *always* changes sign."

"Well, I am sure that is all very interesting," said Dorothy in a voice singularly lacking in conviction, "but does it really matter very much whether these amplitudes change sign?"

"DOES IT MATTER?" roared the Lion in a voice that made his whole audience start up in alarm. "Does it matter?" he repeated rather more moderately. "Of course it matters. The Falsbadour was right enough about that. It is *particularly* important for fermions, as you may soon see. Fermions *always* change the sign of their amplitudes when a pair of them exchange one with another, but if those two were in the same state, then their exchanging really would not have made any difference at all, so the amplitudes would have to be unchanged. Now an amplitude can remain unchanged after it has changed its sign only if it is *zero*. In other words, there is no such amplitude. You *cannot* have amplitudes that have two fermions in the same state. That is Pauli's Exclusion Principle, and you can see how important that is!" The Lion paused and beamed expectantly at Dorothy.

STATES AND THE PAULI PRINCIPLE

The distinction between states and amplitudes is not easy to make. A *state* means, in effect, what the particles can be "doing." An *amplitude* gives the best possible description of the state.

The Pauli Principle then says that no two identical fermions—no two electrons, say—can be "doing exactly the same thing."

"Well, yes," said Dorothy obligingly. "There is just one thing, though." "And what is that?"

"What is a state?" asked Dorothy plaintively.

"Ah. Hmm. Yes," responded the Lion, rather taken aback and for once almost indecisive. He very briefly produced a book. From her quick glimpse of it, Dorothy thought the title to be something like *In a Right State*, but apparently he decided that the time was not appropriate. The girl heard him murmur something about "never can find the right page anyway," though in the case of the Lion, a murmur was like a roll of thunder. He tucked the book away again almost immediately.

"Right," he said as his natural confidence once more reasserted itself. "Well, a state is what it says it is really. It is that state or condition in which a particle finds itself. In particular, when a particle such as an electron is trapped within an atom, then it has discrete energy states. I had better explain that," he added hastily, before Dorothy could ask a question.

He stalked over to the Falsbadour, who was still sitting and strumming idly at his lute, and, with a cursory word of what might even have been apology, lifted the stringed instrument from the despondent minstrel's hand. He gently extended one large, dark claw and plucked at a string, producing a low note. "This string gives a frequency that depends on the length of the string. The same string can give different notes, though only from a definite set."

With another claw, he briefly touched the center of the string as he plucked it again. This time the note was much higher,

R Gilmore

being a harmonic of the first. "So there you are," he said, with a rather inappropriate air of finality, although Dorothy was not at all sure that she *was* there or indeed was *anywhere* in particular.

"Perhaps I can help," remarked the Tin Geek, lurching forward and switching on his chest display. It showed a pair of vertical lines, one on either side and joined by a wavy line across the screen that looked very much like the waves the Geek had shown earlier.

"You see a wave tied down at either end. This shows how the amplitude of an electron or other particle may vary between two points beyond which it cannot go, perhaps because there is a wall of some sort. There is no amplitude at the end points, for the amplitude may not continue beyond them. This limits the options between. The simplest case has one point of maximum vibration—one peak in the amplitude—before the 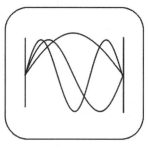 amplitude falls off toward the fixed point at the other end. This case has the lowest frequency possible. Next you might fit in two peaks between the ends, which gives you a wave of half the wavelength and double the frequency. The next possibility is three peaks. And so on. You have a sequence of possibilities, each of higher frequency."

"That's right," broke in the Lion. "If you remember that frequency is proportional to energy, then you can see that when particles are shut into a small space, their energies must choose from a set of distinct values. Particles that are totally free to wander where they will are also free to have any energy that they like. Trap them in a small region, however, and they can only have one of the set of energies that is right for that region."

"Talking of traps," said the Scarecrow. "This looks to me to be the sort of country where you might expect poachers. If we look carefully, we may find a few traps set."

He set off down the path with his distinctive swinging walk, and Dorothy and her companions followed. They had not gone very far when the Scarecrow swung abruptly off the road. "There you are," he said, pointing to some sort of cylindrical pot buried in the ground. "That is what we are looking for."

"And what is it?" asked Dorothy.

"Why, it is a wild fermion trap." No sooner had he spoken than they found themselves abruptly among a horde of blurred, banded stripes that extended through the woods and past the buried trap. "And there is a passing herd of fermions. You may recognize the typical way that the phase varies for free particles as they move along."

Most of the particle tracks skirted the trap. A few passed over it, but most of these only veered to one side or the other and kept on going. Now and then, however, a particle's path would stop abruptly at the trap, with a brief flash as something bright shot away.

"Most of the fermions escape entirely or at most are elastically scattered," commented the Scarecrow. "Occasionally one is captured and falls into a lower energy level within the trap, where it is thereafter bound. It loses energy in the process. Indeed, the only reason why it *is* bound is that it does not have enough energy to escape and travel freely. Energy has to be conserved, as you know, so the energy that the fermion has lost is carried away by a photon, a 'particle of light.' This trap is the usual steep-sided 'square well' design," the Scarecrow went on. "They make the traps as deep as they can so that the traps can hold as many fermion levels as possible. A level cannot hold a lot of fermions because of that Pauli Principle you have just heard about." Dorothy remembered what the Falsbadour and the Lion had managed to tell her about the Pauli Principle: that only one fermion could exist in any state. "The trap is also made as deep as possible in order to reduce the loss by tunneling when they lift the trap up."

Tunneling was another word that Dorothy had not come across previously, but before she could ask about it, they heard the sounds of furtive footsteps. The Scarecrow hustled his fellow travelers out of sight behind some bushes, just before a motley group of disheveled people appeared from the direction of the path. If they were poachers, as Dorothy presumed them to be from what the Scarecrow had said, then presumably also they were trying to act in secret. Indeed, they did creep along with exaggerated stealth, one or another of them stopping every few moments to hold his finger to his lips. Unfortunately, whenever one did so, the others invariably collided with him, and all fell into the bushes with a great deal of thrashing about and cracking of branches.

This noisy stealth served to slow their advance, but even so, the band of marauders reached the pit eventually and crowded around it. Their approach had been further hampered by the spades and shovels that they carried slung over their shoulders. Dorothy realized for what purpose when the entire group began to dig around the well. Each thrust of a spade produced a peculiar ripping sound. Dorothy commented on this. "Whatever are they digging through?" she whispered.

"The trap was embedded in the fabric of space–time. Very tacky stuff, that."

The band of desperadoes appeared to be quite unaware of this exchange and crouched around the shaft, struggling to lift it onto a higher energy level so that it would be above the general surrounding level. The trap appeared as a deep cylinder, and the gang held it high as they prepared to carry it off. Because they encircled the object, this required some of them to walk backward and others sideways. Initially there was no agreement on this, and they surged to and fro, occasionally rotating in a complete circle. Eventually they reached some level of agreement and moved away, carrying the cylindrical well. Every now and then, as they staggered off, a fermion would pop out through the side wall and rush off through the woods, free once more.

"That is tunneling," said the Lion, answering the question that Dorothy had not been about to ask. "Because the energy of any fermion in one of the levels within the well is less than the height of the walls, a fermion within the wall has negative kinetic energy. Classically, this is simply not allowed. In a quantum state, a particle *can* have negative kinetic energy, though its amplitude falls away quickly. The higher the wall, the more negative is the kinetic energy of a particle within it and the more quickly its amplitude will decrease within the wall. Even so, if the wall is thin, there may still be a significant amplitude on the other side of it, and this gives a probability that a particle that should be trapped inside the well will in fact appear outside, having in effect *tunneled* through the wall. When a potential well contains fermions, the lower levels soon become filled, and fermions that are added later are forced to go into higher levels near the top. There they are not so deeply bound and can penetrate the walls more readily, so that they quickly tunnel out. You can read all about it in my book." The Lion reached into the undergrowth and produced a brightly jacketed book with the title *Tunnel Visions* by Isa C. Lion.

TUNNELING

This is a purely quantum process. In classical physics, particles could not go where they would have a negative kinetic energy. In quantum physics, they can still have an amplitude in such a condition. This falls off quickly with distance but may still give a small but finite probability for the particle to appear on the other side of a *thin* potential barrier.

Tunneling occurs, for example, in the radioactive α decay of uranium nuclei.

"How do you always manage to find a copy of one of your books whenever you want it?" Dorothy queried.

"Oh, there are a lot of them about. They are very widely distributed. I can usually find a book, though not always the page I want," answered the Lion evasively.

While they had been talking, the gang of poachers had vanished from sight, and even their whispered cries of "Stop pushing" and "Left, left. I said go *left*. Oh sorry, I meant right" had faded into the distance. It appeared that nothing further was going to happen where they were, so the companions continued on their way through the woods. Occasionally they passed other holes. Some were deep like the fermion trap, and others were wider and quite shallow.

"Those are probably boson traps," remarked the Scarecrow. They can hold a lot of particles without having to be so deep, because all the bosons will go into the same level."

"Why is that?" asked Dorothy.

"It is because there is no Exclusion Principle for bosons," said the Lion, confidently. "There is no reason why two bosons should not go into the same state, so naturally they prefer to."

"I do not see why that should be," replied Dorothy. "Even if they *can* go into the same state, is there any particular reason why they *should?*"

"Well yes, there is, actually. You remember that the amplitude for a particle describes that particle as well as it can be described and gives the relative amount of its presence in any state or condition. An amplitude can describe as many particles as are present. If two particles are in the same state, then the amplitude for that state is doubled, compared with the case where only one particle was in that particular state. As you now know, the probability for particles being in any specific state is given by the size of the amplitude squared, which is to say that it is multiplied by itself. Accordingly, twice the amplitude will give *four* times the probability for only *two* particles to be in the same state. For ten particles you would have a hundred times the probability—ten times as much as if each of the ten particles were in a state with its own different amplitude. For larger numbers of particles, the state becomes more and more probable. In effect, the more particles that are present already, the more effectively the state sucks particles in. This is known as Bose Condensation."

They had by now made their way back to the Building Block Road. As they strolled along, Dorothy talked to her new friends about all the things that had happened to her recently, including how she had come from Kansas and had been greeted by the Witch of Mass. "She told me there was another Witch who was called EM. I live with my Aunt Em. I wonder if the witch is anything like her." Dorothy was beginning to miss her Aunt Em and thought fondly of her kindly, careworn face and her grey hair (now with a permanent wave in celebration of her visit to the city).

Her reverie was abruptly interrupted by a roar of thunder and a great flash of lightning that seemed to strike right by the roadside. Dazed by the sound and half-blinded by the brilliant light, Dorothy shook her head to clear it. As her vision returned to normal, she became aware of a figure nearby where no figure had been visible a moment before. Now there towered before them a slender woman in a long, dark robe. On closer examination she did not in fact appear to be especially tall, but nevertheless she towered. As one who could both operate within the scope of an atom and extend her reach over entire galaxies, her apparent size was largely a matter of social convenience. She could tower as much as she wished.

Dorothy saw before her a pale face with high cheekbones and hooded eyes below eyebrows like dark thunderbolts. In her hand this Witch, for Witch she was, carried a staff made of captive lightning that hummed and crackled. The staff was surmounted by a large, round ball that glowed with a blue light and emitted short, fat sparks. The whole figure was outlined in a similar blue corona. Her wide, mobile mouth twitched with irony as she addressed Dorothy.

"Child, I am EM!"

R Gilmore

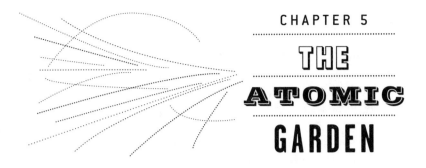

CHAPTER 5

THE

ATOMIC

GARDEN

Dorothy and her companions stared amazed at the sudden and unexpected appearance of the Witch of Charge. Truth to tell, only three of them were staring. The Tin Geek was not staring at anything, because the nearby lightning discharge had crashed his system again. The Scarecrow noticed this and quickly inserted his system disk into "drive A," as invited by the screen message. The Tin Geek recovered with some initial confusion, and then he also stared at EM, who was watching him quizzically with one eyebrow raised.

Dorothy felt a bit scared, but above all she felt angry.

"You should be ashamed of yourself," she said, "picking on the Geek just because he is the only one of us who is affected by electricity."

"You are totally mistaken," replied EM. "You are each and every one of you affected by electricity and at a very deep level. Everything around you in your world is essentially electrical. If electrical forces were withdrawn the effects would be—unfortunate. Observe."

She pointed a long finger at a pebble by the roadside. It promptly exploded like a hand grenade with a shattering bang. "That," she said, "is what would happen to you in your world if electrical interactions ceased. When electrons are confined within the small space of an atom, they have a large momentum because they are so constrained. If suddenly there were no electrical attraction to keep them bound, they would fly apart with the corresponding velocity, which would be quite as much as in a more conventional explosion.[1]

"My domain extends over the whole world and everything that is in it," continued the Witch. From somewhere about her person, she drew a fuzzy ball that she held up and then tossed toward them. As it came, it shrank and multiplied, shrank and multiplied, to become a myriad of tiny objects and finally a mist of particles too small to be distinguished. When this mist reached them, it melted into Dorothy's body, blending with and finally becoming the atoms of which her body was made.

"Every atom in your body is mine. You are mine because you are made of atoms, and these atoms are dependent above all else on the electromagnetic interaction. Everything you see about you depends on it. Indeed, the very fact that you *can* see depends on me. The photon, light itself, is the basis of the electromagnetic interaction.

"*I am light!*" she cried exultantly.

Abruptly there was a blaze of light all about them. As before, they could see nothing for some moments after their eyes had been blinded by the intense radiance. When they could see again, there was no sign of their recent visitor. The Witch had vanished.

They walked on, but not far beyond the spot where they had met EM, they came across a notice by the roadside.

Garden produce
from the Atomic Garden
now available

any and every object
in the whole wide world

Intrigued by this notice, they turned aside and entered the garden. A scene of rich profusion greeted them. Crowded beds, occupied by every variety of atom, surrounded them. There was simple Hydrogen in one bed; another contained every variety of Rare Earth. Oxygen gave a refreshing atmosphere, and nearby Carbon hinted at inconceivable complexities. Behind it all was the heavy, dangerous scent of Uranium.

Nearby was a stooped figure, seemingly half as old as time, who was pottering about on a path between the beds. He turned and straightened up marginally as they approached.

---

[1] It is of course quite impossible to suppress the electrical interaction in this way. If it were possible, it would certainly be dangerous. DO NOT TRY THIS AT HOME.

"Arr, do you be come about some of they garden products?" he asked.

"What sort of products do you have?" asked Dorothy.

"All sorts. Any rates, we do be having all sorts of atoms, and you can make whatever you likes from they."

"Can you tell me anything about your atoms, then?" asked Dorothy, just to be perfectly clear what was going on.

"Aye, to be sure. Just you come along with me. I be about to begin a new batch of Nitrogen."

Dorothy and her companions followed him as he doddered off down the path. He stopped by a blank area, reached into a bag he was carrying, and placed something in the center of the plot. It was not a blank area of soil, the girl noted, but a *completely* blank area. Whatever it was he had set down was apparently isolated in a region of empty space. He repeated the action several times. Dorothy couldn't actually see anything, but she was sure he had put something there. She asked him about it.

"Why they be nuclei for the atoms. You needs to start with a nucleus if you wants an atom. As my dear old granny says,

> *When a nucleus is planted,*
> *You'll get the atom that you wanted.*

All I needs do now is feed they with some electrons."

He lifted up a sort of can from the path nearby and tilted it over the region he had seeded. A hazy shower fell from the can, and a fuzzy ball developed around each nucleus that he had planted there. Initially perfectly spherical, as they grew larger they became more complicated, with outlying lobes. They remained pretty fuzzy and featureless, however.

"There you be. A rare marvel of intricate structure, you mun agree."

Dorothy didn't really. The atoms still looked fuzzy to her, and she said so. The Gardener looked surprised for a moment, and then his wizened features cleared.

"Ah! I bet that you was looking at they in *space*. Of course they will look fuzzy to you in *space*. You must look at they in *energy*, girl. There is where they looks sharp to you."

"How do I do that?" asked Dorothy. "I always look at the position of things. It's the only way my eyes *can* see. How can I possibly see energy?"

"A poor sort of eyes those. Perhaps, young though you be, you needs a pair of they spectacles. Here, have a borrow of a pair of mine."

Dorothy took the glasses he offered and put them on. They were small half-glasses, and she strongly suspected that they did nothing for her image. They did change the way she saw things, though. Where the Gardener had planted his nuclei, she saw a sort of deep funnel, tapering down to a point and spreading out as it rose, like the mouth of a trumpet. She could see similar trumpet shapes farther away, and in them she discerned vague shapes that she took to be electrons. She commented on this, and the Lion, who had been silent for a remarkably long time, now hastened to enlighten her. As always, his voice boomed out confidently.

"That is the potential well created by the positive electric charge on the nucleus. The negatively charged electrons are attracted to the nucleus, and their potential energy is less the closer they are to the nuclear charge. You have seen how gravity gives objects that are high above the ground a *potential energy* that they may convert to kinetic energy as they fall and fall ever the more quickly. The electric charge of the nucles will attract electrons in the same way as the mass of the Earth would attract you if you were so incautious as to fall into a well, and you would gain speed and kinetic energy until you hit the water at the bottom. In effect, it is as though the electrons fall down a well as they approach the nucleus, and it is known as a *potential* well. This well captures electrons and holds them around the nucleus, as you see."

"Why don't they just fall right down to the bottom of this well?" asked Dorothy reasonably.

"They are too *light* to fall any farther. When electrons fall they get closer to the nucleus, and if they are definitely near the nucleus this means their *positions* become better defined. Because of the Heisenberg relation, this means the spread of the electron *momentum* becomes greater. You cannot have a momentum *spread* without having momentum, so the electrons must gain more momentum. The momentum they gain comes from the kinetic energy they get by falling down the well, but there is not enough of this to allow them to be too compressed, and they end up trapped as the sort of electron cloud that you see. They do not have enough energy to get any smaller, but the electric field stops them from spreading farther out. Atoms are the size they are because that size happens to be as far as the electric attraction provided by the nucleus is able to compress the electron cloud."

As the Lion was talking, the Gardener had begun to pour electrons over the new nuclei. One of them fell into the potential well that Dorothy was watching and took its place in the lowest energy level. A second electron followed, and it also ended up in that same level.

"That is strange" remarked Dorothy. "Actually everything here is strange, but I have been particularly told that only one electron can fit into a state. Now I see that two have gone into in the *same* level. How is that?"

"Ah, 'tis because of the spin, my dear. Electrons can have spins in two different directions, so you can put two of they in a level. As my old dear granny always says,

> *Two electrons fit in,*
> *Because of their spin.*

And surely 'tis always so."

Dorothy looked beseechingly at the Lion, who hastened to her rescue.

"Let me say a few words about electron spin. I could say far more than a few, of course," he added, producing a book that had apparently been concealed in one of the garden beds. It was titled *Do Electrons Get in a Spin?* by Isa C. Lion. "I suppose, though, that a few words are all that are needed at the moment." Regretfully, he put the book away again. "Electrons and all fermions have spin; they behave rather like little tops that spin around all the time. They cannot stop this spinning, and all electrons have the same spin—the same *angular momentum*. Angular momentum is a bit like the momentum you have met already except that momentum previously went with movement in a line, whereas this momentum has to do with rotation. In a quantum state, the angular momentum can change only by an amount equal to $\hbar$, which is usually called *one unit* of angular momentum. The electron may take only one of the two values $+\frac{1}{2}\hbar$ and $-\frac{1}{2}\hbar$. This is usually referred to as 'spin up' or 'spin down.' These two values of spin correspond strictly to different quantum states, so you can have twice as many electrons as the Pauli Principle might otherwise have suggested. It is quite simple really," he added unconvincingly.

Dorothy now saw another electron fall into the atom she was observing. This one fell down the well after the others, but stopped at a level much higher up.

"He must go to a higher level. There was no more room for he in the one below. As my dear old granny says,

> When a level is quite full,
> Must go higher, that's the rule.

There be plenty of levels up there for they to go to."

Another electron fell in and settled in the same level. Dorothy assumed this to be because of the spin states that she had just heard about because she had been told that two electrons would fit into each level. Another electron dropped in, and she expected, on the basis of what she had seen so far, that this would go into a noticeably higher level. Instead it dropped down to what appeared to be much the same energy level as the previous two.

"What's this?" she exclaimed. "First you said that only one electron could go in a state. Then you said that there could be two, because of the different spins."

"That is quite true," interrupted the Lion swiftly. "The electron has two different spin states. These both have the same energy in the electric field of the nucleus, so you get two electrons in the same energy level. There is absolutely no deception." He finished confidently.

"Perhaps not," admitted Dorothy, "but *now* there is a third electron in the same level. What do you have to say about that?"

Perhaps not surprisingly, the Lion *did* have something to say. "That is different. Well, actually it is rather similar in a way, but it is still different. As well as spinning within themselves like little tops, the electrons can also rotate around the nucleus, a bit like planets in orbit around the sun."

"If the electrons are attracted to the nucleus like planets to the Sun, aren't they all in orbits rotating about it?" interrupted Dorothy.

"No, it is not like that at all. Planets do not fall into the Sun because of the centrifugal force caused by their rotation." The Lion paused briefly and was obviously considering producing a book he had written about the mechanics of the solar system, but he thought better of it and continued. "Electrons do not fall into the nucleus because, as you have seen, it would require more energy to localize them near the nucleus than they can gain from the electrical potential when they fall. Some electrons have a rotation about the nucleus, but some, including the states of lowest energy, do not.

"For those that do, the rotation will give angular momentum, and this gives more states. There are in fact three for the lowest amount of angular momentum. Each of these 'orbital angular momentum' states can have two electron spin states, of course, making six in all. They all have much the same energy, so when you add in the two states that you get for electrons that do *not* have any orbital motion, that makes eight in all. It does get a bit complicated," he conceded finally. "Perhaps it will be better if you simply accept that there are a number of possible energy levels available each of which can hold only a certain known number of electrons.

"As each electron is added to the atom, it cancels or *screens* one unit of the nucleus's positive charge, and any incoming electrons will see a smaller overall attractive charge. When the number of electrons in the atom is equal to the charge on the nucleus, you have a *neutral atom* that does not readily gain more electrons."

As Dorothy watched, the atom captured more electrons from the Gardener's spray until there were five in the higher level. With each electron captured, the potential well shrank, until finally there was little sign of it away from the immediate vicinity of the captured electrons. Now any further electrons just drifted past without stopping.

Because nothing more seemed likely to happen to the atom that she had been watching, the girl looked over the top of her "energy glasses" and glanced around the garden. She noticed bright objects flashing rapidly around and darting from atom to atom, rather like hyperactive bees, she thought. Often they would just glance off and continue on their way, but occasionally one would collide with an atom and vanish. A bright shape flashed by in front of her and collided with the very atom she had been examining. She quickly glanced down through the lenses of her spectacles, and again she viewed the energy levels. Now she saw that there was only one electron in the lowest level, not two as there had been. As though to compensate, there was now an electron in a higher level where none had been before.

"There you see that a photon has given its energy to an electron and excited it to a higher energy level. That is something that I saw frequently when I was in my field," remarked the Scarecrow from just behind her. "As I mentioned before, the photon is a boson, and bosons are not necessarily *conserved.* Unlike a fermion, a boson may simply cease to exist and give up all its energy to an electron in the atom. This raises the electron from its initial energy level to a higher one, as you have seen happen here. It has consequently left a *hole,* one less electron in the lower level than that level is capable of holding."

R Gilmore

As he was speaking, Dorothy saw the electron in the higher level fade from her view and noted that another electron appeared in the unfilled outer level. As far as she could tell, the two events happened at the same time, and, also at the same time, she saw a brief flash as a bright photon sped away. Soon after this she saw repeated a similar set of events, but this time ending with the lowest level filled and only five electrons in the outer level, so that the atom appeared as it had been before the impact of the initial photon.

"The electron has fallen back to the hole in the lowest level, and the energy it lost in the process has been carried away by photons. Photons are very good at carrying off energy, for they may readily be created so long as there is electric charge involved, and electrons are well equipped with charge. Electrons—and anything else, for that matter—always tend to fall to lower energy levels if there is nothing to stop them. You yourself would tend to fall to the lower energy level if there were a hole beneath *you*. Usually electrons in the outer levels cannot fall because electrons in the filled lower level get in their way and block them. Normally you do not fall because the ground under your feet gets in your way and blocks you. Indeed, it is the electrons in the ground that are blocking the electrons in your feet. The electrons in an excited atom may fall down in several stages, but the atom eventually returns to its lowest energy, in what is known as its *ground state*. I have spent so long up on a post staring at the ground that I am something of an authority on ground states," added the Scarecrow.

Hearing the word *authority* applied to someone else was too much for the Lion, who immediately hastened to get into the conversation. "The photons that may be emitted give the atomic spectrum for light emitted by that type of atom. This spectrum is a unique set of frequencies corresponding to the energies that photons have been given by electrons that have jumped between the particular energy levels of the atom. This set of frequencies would show up as a series of colored lines if the light were passed through a spectrometer, and this set of lines identifies the atom as surely as the lines in your fingerprints would serve to identify you. I have explained it all here. We can look up the relevant sections later," he added, lifting a stone and producing from under it a book that was titled *Light-Fingered Atoms?* by Isa C. Lion.

ATOMIC SPECTRA

The electrons in an atom are located in distinct levels that have specific energies typical of that atom. When an electron changes from a higher level to a lower level, it emits a photon that carries away the surplus energy. The photon energy is simply the difference between the energies of the initial and final electron states, and this energy gives the photon a specific and distinctive frequency. The light emitted by atoms of any type is made up of such photons, and it identifies the atom uniquely.

"Though spectra can tell you a lot about an atom, the most interesting thing about atoms is of course that everything in your world is made of them. The number of electrons that there are in the outer, or *valence*, level of an atom determines its chemical properties," boomed the Lion, putting away the book as he eagerly grasped the opportunity to move on to a new topic. "As my dear old granny do say," cut in the Gardener,

> *When an atom holds what he is able,*
> *That determines his position in Mendeleev's Periodic Table.*

Dorothy and her companions all looked at him.

"Well, she don't say it *very often*," he admitted.

"It is quite true, though," conceded the Lion, taking this couplet as an occasion for a further display of knowledge on his part. "Atoms are the smallest bits into which the *chemical elements* may be divided, and everything in the world around you is made from atoms that have combined into *molecules*.

The molecules are the smallest portions of the various *chemical compounds* of which everything in your home world is made, including yourself."

"And you too, surely?" said Dorothy, seeing no reason why she should be singled out as an example of chemical composition.

"Well no, actually. I am not. I am an allegorical construct, you see, and so purely fictitious. So are you when you are here. You are currently much too small relative to the atoms here, but back home your body is made up of enormous numbers of them.

"Anyhow," he continued, returning abruptly to his favorite occupation of explaining things, "the way that any atom combines with others depends on the number of electrons in its outer shell, which is what chemists call the highest energy level that contains any electrons. Its chemical activity depends on how close this outer level is to being *full*—that is—to holding as many electrons as it can.

"You will have noted," he went on, "that there is quite a big difference in the energy of the different levels. When the last electron to be added is forced into a new level, its energy will be a good bit more than that of its companions in their lower levels. If there is another atom nearby, and this atom has one gap left in *its* outer level, then the total energy might be reduced if the lone electron switched over to the second atom. Now I know what you are going to say." The Lion held up an enormous paw to silence a protest that no one seemed inclined to make.

"You will argue that the other atom will already have its full complement of electrons—that it will consequently be electrically neutral and so not attract any more electrons." Dorothy couldn't hear anyone arguing, apart from the Lion himself, but he seemed perfectly capable of presenting both sides of a conversation. "This would indeed be the case for electrons a long way away, but any electron that was close in—that was actually in the outer filled level— would not be screened from the nuclear charge. It would be in just the same situation as the other electrons in that outer level. In such a situation, the atom has more electrons than needed to balance the positive charge on its nucleus and is known as an *ion*. The extra electron is not too firmly attached, but while it is there, the ion has a net negative charge because it has one more electron than it needs to balance the positive charge on its nucleus. It is a negative ion. The atom that gave up the electron in order to empty its outer level is correspondingly positively charged. It is a positive ion.

Because ions of a given type have the same overall electric charge, they *repel* one another and tend to run away. You could say that they are *cowardly ions*." The Lion chuckled appreciatively. No one else did, however, so he went on quickly. "A positive and a negative ion have opposite electric charges, and so they attract each other. This attraction gives an *ionic bond* and results in the two atoms combining to form a molecule. Such molecules are pretty simple affairs. Salt is a good example. A molecule of salt is simply one positive sodium ion combined with one negative chlorine ion. It contains only two atoms, and that's it.

"More complicated molecules can result from *covalent bonds*, in which electrons are *shared* between atoms. This is most striking in the case of carbon."

CHEMISTRY AND THE PERIODIC TABLE

The number of *valence electrons* in its outermost energy level determines the chemical behavior of an atom. There may be one electron short of a filled level in one atom and, in another, an electron on its own in a higher level. Energy is released if the lone electron transfers to fill the space remaining in the other atom, and the resultant electrical force binds the two atoms together to give a molecule, the smallest part of a chemical *compound*.

The Lion's smooth and almost hypnotic speech was suddenly interrupted at this point by a measured chanting.

> *Hail great Carbon, blessed be thy bonds.*
> *Hail to thee, four-valenced one, the builder of life.*

The four companions turned abruptly to find that a group of people had approached while the Lion was talking. They were dressed in simple, smock-like garments and had considerable facial hair combined with expressions of such earnest sincerity as to appear almost simple-minded. One of the group, who seemed to be their leader or at least their spokesman, addressed Dorothy and her friends.

"We are members of the Carbon Commune. We are dedicated to the task of crafting molecules in all the manifold varieties that may be constructed with the wondrous carbon atom as their base. All of our work is completely organic, of course.[2] Carbon is an atom in perfect balance. It has four valence

---

[2] Chemical elements that contain carbon are known as organic compounds. Those that do not are said to be inorganic. This nomenclature arose because life on Earth is based on the amazing variety of compounds that may be made from carbon.

electrons in its highest energy level, or outer shell, as we call it. For completion it would require eight. Thus you see that it at the same time *lacks* four electrons and has four *in excess*. Should it gain four electrons from other atoms or should it lose the four it has? Either method would serve just as well to give a completed level, and either is possible. Because both methods are possible, both are compulsory, so carbon does *both*. Carbon will graciously *share* its four electrons with four in other atoms so that, in this condition, it combines an amplitude that has eight electrons in its outer shell with an amplitude that has none. Both choices give a completed outer shell for the carbon, because having no electrons in the erstwhile outer shell means that the 'outer shell' is now the complete one just below. Either way, this sharing means that the atoms are attracted to one another,[3] and so the overall effect of this 'exchange interaction' or 'covalent bond' is attractive. Come with us and see what sort of molecules we make."

The troop of commune members marched off with a patter of sandal-clad feet and led the way through the garden back onto the Building Block

[3] This is by no means self-evident, but it works out to be the case.

Road. At this point, most of the blocks in the road bore the symbol C or H, with an occasional O or N. They had gone but a short way down the road when they came to a large barn-like building. They were led into a lofty hall where many more members of the commune, similarly attired in smocks, were working away assembling complex structures. At first sight, these seemed to be made from little balls connected by short rods, but on closer examination, the balls had the rather fuzzy look of the atoms that Dorothy had seen already and were connected by hazy blurs that glued them together where they touched. The first person they came to was young, an adolescent with no beard or moustache. He was putting together a rudimentary-looking structure that was nothing more than a single central carbon atom with four hydrogen atoms spaced around it.

"This is Billy. He has just joined the commune, and so we are starting him with the simplest organic molecule, methane. This molecule shows clearly that carbon can have four bonds, one to each of the four surrounding hydrogen atoms. The four electrons in the outer shell of carbon are sharing the single electrons from each of the four hydrogen atoms, as you see."

They moved on. The next item under construction was a bit more complicated, with two carbon atoms bonded to each other and the pair surrounded by six hydrogen atoms. Five were connected directly to carbon atoms, but one had a new type of atom in between.

"That extra atom is an oxygen atom. Organic molecules often contain oxygen as well as carbon and hydrogen. The bigger molecules can contain all sorts of other atoms as well. It is easier to understand the nature of an organic molecule if you see a diagram that shows how the atoms are arranged." The Tin Geek's chest screen lit up, and he displayed a little picture of the molecule that Dorothy had just seen, with the types of atoms shown in different colors.

As Dorothy followed her guide around the workshop of the commune, her electronic companion showed a range of molecules of steadily increasing complexity. Or rather, Dorothy thought to herself, the molecules that she was shown were of steadily increasing complexity, and the Geek showed her a diagram of each one, but in between she noted that he displayed again and again the molecule he had started with. They had reached a section of the building where several groups of commune members were working together on a vast molecule, a construction that looked like a great double helix, extending away into the dim shadows at the far end of the building. At this point the Tin Geek's articulated knee joints suddenly gave way, and he sank to the floor with a loud clang.

"Do not worry," he said, "jusht a little unshteady, thatsh all. I'll jusht go into relaxation mode for a couple of minutes and I'll be OK." On his display

R Gilmore

screen, which was currently crammed full of the atoms that made up the last molecule he had been showing, there appeared a little round shape that was essentially one large, hungry mouth. This moved at random through the crowded picture, gobbling up the image and leaving a blank trail behind.

"What has happened to him?" exclaimed Dorothy in some dismay. "This is not like him at all!"

"I am afraid that our friend has been engaged in a little quiet tippling," answered the Scarecrow. "I do not know whether you noticed the molecules that kept appearing on his display: They were ethyl alcohol. I am afraid that our cybernetic companion is cyber-drunk."

"How can that possibly be?" asked the girl. "All he has done is show pictures of the molecules. *That* could scarcely make a person drunk!"

"Not a person, perhaps," murmured the Scarecrow, "but you must admit that he is not exactly a *usual* sort of person. He is really more of a computer, and programs rather than food are his input. So you see, the *formula* of alcohol is quite enough to make him a little inebriated. We had better get him out of here."

Together Dorothy and the Scarecrow loaded their friend onto the Lion's back and retraced their steps along the Building Block Road toward the garden. Despite the effort of preventing the Tin Geek from falling off the Lion's back in a tangle of uncoordinated metal limbs, Dorothy had time to notice that many of the blocks with their different symbols were fused together into large, complex assemblies. Throughout the journey, the Geek kept up a sort of manic electronic beeping noise and waved his arms about. As they got farther from the influence of the carbon compounds and re-entered the garden, he quieted

R Gilmore

down and was soon able to stand, if a little uncertainly, on his own feet.

"So many substances, and all depend on whether one particular type of atom is present. It seems quite remarkable," remarked Dorothy, who felt that it was not right to comment further on her comrade's behavior. "If atoms are so different, what decides this difference?"

"Why, it is the number of electrons in the outer shell, nothing else. The number of electrons in the atom determines its chemical behavior."

"So why *do* some atoms have more electrons than others?"

"Why, 'tis all 'cause of the charge on their nucleus, my dear." The Gardener had joined them when they re-entered the garden but had been silent up until now. "An atom has just the electrons it do need to match the charge on its nucleus. 'Tis as my dear old granny says,

> *All the electrons an atom do need,*
> *Is the number to match its nuclear seed.*

The answer, my girl, lies in the nucleus."

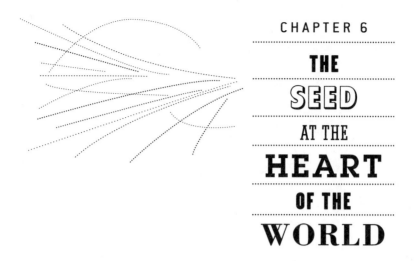

CHAPTER 6

THE

SEED

AT THE

HEART

OF THE

WORLD

"What and where is this nucleus then?" asked Dorothy.

"Why here it be, at the heart of the atom," replied the Gardener, pluck-ing out a nearby atom that happened to be carbon. He held this out for Dorothy to examine. She looked closely at the diffuse object, but could not see anything different in the center.

"I don't see it," she said.

"Well, certainly, it do be very small," responded the Gardener. "Here, let me make him grow a might." The blurred shape of the atom began to ex-pand. It grew and it grew beyond all reason until it spread far on either side, being now about the overall dimensions of a football stadium.

"How did you do that?" gasped Dorothy. The Gardener smiled.

"A fine gardener I be if I can't make things grow! Now you look and see if you can but spot the nucleus where he do rest in the middle."

Dorothy looked. She saw the misty cloud of the electron probability dis-tribution spreading far on either side. She searched vainly over the foggy re-gion before her, but she still could see nothing in all the great featureless ex-panse of electron amplitudes. She said as much to the Gardener.

"Why no, that's as may be. Even though ye cannot see she, she be there right enough. That nucleus still be mighty small though, and 'tis hard to say where she be. The nucleus do be regular lost amongst all that electron cloud. 'Tis as my dear old granny do say,

> Even though you do expect uns,
> You can't see the nucleus for the electrons.

If we are to find she, we need a little help. By good luck the local group of alpha scouts are holding their 'MeV[1] a job' week this week, and there be a gaggle of the little tykes just arrived."

Dorothy looked behind the Gardener as he spoke and could see, or at least was somehow aware of, a crowd of small shapes. They were rushing about in all directions, at high speed, and emitting shrill, excited noises. Under the girl's fascinated gaze, the Gardener went patiently about the job of carefully shepherding this flock of energetic helpers and sending them on their way across the great expanse of the atom. The alphas went rushing out with high delight, following random paths through the region. Most rushed straight across with no apparent deviation. Some veered around the mid-point of their path and came rushing out in a direction slightly different from the way they had gone in. Most of these were deflected only slightly, and Dorothy did not at first notice the deviation, but then one veered at such an angle that it could not escape her attention.

"Why are some of them swerving aside as they charge across?" she asked no one in particular. She was not in the least surprised that it was the Lion who chose to answer.

"It is because of their charge: their electric charge. The alpha particles have a positive electric charge, and so does the nucleus. As a consequence they repel each other; the alphas are pushed away from the nucleus and swerve aside. The closer they come to the charge on the nucleus, the stronger this repulsion is and the more they are deflected. Some of them come very close indeed. Just look at that one."

Dorothy saw then that among the various alphas that had rushed across with very little, if any, change in their direction, there was one that suddenly halted and rebounded almost directly back along its path.

"When you consider how much energy those alphas have, you can appreciate that it requires a very powerful force to turn them around like that. Electric forces get stronger the closer the charges that interact are to one another. In order to get a strong enough force, the alpha must pass *very* close

---

[1] An MeV (pronounced "em-ee-vee") is a measure of the energy a particle possesses. When Ernest Rutherford performed the scattering experiments that revealed the existence of the nucleus, he used alpha particles with energies of a couple of MeV.

to the nuclear charge. This shows you that the nucleus must be very small indeed."

"Excuse me," Dorothy returned. "I do not see how it shows that at all."

"Oh, but it does," interrupted the Scarecrow. "For example, just think how difficult it would be for you to get within a few feet of the whole Earth."

"It's not difficult at all!" the girl retorted. She was becoming rather tired of what she felt to be inappropriate analogies. "I am within a few feet of the earth most of the time."

"No, no," countered the Scarecrow. "I don't mean earth like you get on the ground. I mean Earth, the planet. I know all about earth," he added; "I have been staring at it for years. You may be able to get close to the *surface* of the Earth, but you cannot get close to *all* of it. You may be close to the Kansas bit of the Earth, but you are a long way from Australia. You cannot get very close to most of the Earth because the Earth itself is so large and spread out. You will always be a long way from most of it. The only way you could get close to all of it would be if it were small and each bit of it close to the rest. In order to get the sort of force that is needed to turn those alphas around, they must get close to *all* the nuclear charge, and this can happen only if the nucleus is itself very small. If we try to find the place the alphas were running away from, we may find the nucleus. Let's see if we can."

They walked across to the spot that was, as far as they could judge, the place that all the charged alphas had been avoiding. They searched around diligently, and suddenly Dorothy saw it. Floating in the fog of the electron amplitude she saw a tiny, dense object about the size of a grain of rice.

"Is that it?" she exclaimed in disbelief, reaching out to pick it up and examine it more closely. To her surprise it wouldn't budge. Small and insignificant-looking as it was, she could no more move it than if it had been the head of a nail driven into solid granite. "Why won't it move?" she exclaimed in frustration. It had been more of an involuntary expostulation than a deliberate question, but that was all it took to prompt an answer from the Lion.

"It will not readily move because it is so heavy. The nucleus, small as it is, contains over 99.9 percent of the mass of an atom. It is as though the mass of a very large building were concentrated in one brick. That would not be an easy brick to lay."

"I still cannot believe that something so small should be so heavy. Anything so heavy is bound to be pretty big, in my experience. It is only common sense."

"That is the problem, of course," returned the Lion with a slight sigh. "Your experience does not extend to this small scale. Your common sense is fine in common situations, situations of which you have already had experience, but it is no guide to the unfamiliar. *Here* it is small objects that will be heavy, because only heavy objects can be small. If something has great mass, then it can have a large momentum without too much movement and so can be well localized and small. Light things have no choice but to be large because they are so spread out. You expect large things to be heavy because all the objects you normally experience have actually been collections of *large* numbers of *small* objects, namely atoms. Naturally, the more objects you have, the heavier they are, but you have no experience of the relative weight of large and small things on the individual-particle level."

"Perhaps that is all so," said Dorothy grudgingly, unwilling to abandon too readily her intuitive understanding of things. She pushed angrily at the tiny form of the nucleus. "Look!" she cried, "I do believe that it is beginning to move slightly."

"Excuse me, Ma'am," said a new voice from just behind her. She looked around and saw a large, neatly attired man. He had regular if inexpressive features, short hair, and a long white coat with a row of pens and pencils in the breast pocket. He reached past her and deftly picked up the nucleus.

"How did you do that?" asked Dorothy in a voice tinged with surprise and a certain feeling of outrage that he had been able to do with so little apparent effort what she had been unable to accomplish.

R Gilmore

"It's what I have been trained to do, Ma'am. I am an Inspector of Nuclear Weights and Measures, and it is my job to check on the parameters of sample nuclei—their sizes, masses, and other similar properties." He said this in an unemotional way, as though he were reading from an invisible manual. He then took out from an inside pocket a small plastic case, such as might contain a set of fine screwdrivers. When he opened it, Dorothy could just recognize the largest item in the set as a tiny metal ruler. The contents of the case were arranged in decreasing size and he took out one of the smallest objects. Next he inserted a jeweler's eyeglass into one eye, screwed up the other, and measured the size of the nucleus.

"Hmm, slightly under five fermis.[2] OK, that's about right for this nucleus, so long as this is the atom I think it is. I had better make sure." He opened a canvas bag that he had been carrying and took out a square metal box with a number of small windows scattered around it. He flicked a switch on the side of this box. The overall mist of electron amplitudes that surrounded them condensed briefly into six clumps spread randomly around the area occupied by the atom. These quickly spread together until once again there was an apparently featureless, stadium-sized cloud surrounding them. Abruptly this again contracted into six separate regions, though their positions bore no obvious resemblance to the six seen previously. As before, these six amalgamated into a uniform distribution that lasted until it in turn solidified to six separate packets. The whole process was repeated several times, until the Inspector flicked the switch off again.

"What was happening?" asked the girl curiously.

"That is a *strobservoscope*," the Inspector responded in a calm and informative manner. "You will not have seen one of these before.[3] You may well have seen a stroboscope, however. That is a device that gives out bursts of light one after another, so that you may see things in quick successive flashes. This device is similar in that it makes a series of observations of the positions of electrons in the atom. Each observation localizes the electrons at the positions observed, though their amplitudes soon spread out."

The display had been entertaining, but Dorothy didn't really see what it proved. The successive observations had shown the electrons in completely different positions, and she did not think that any overall pattern had been evident. "The different observations had nothing in common," she pointed out.

[2] A fermi is a measure of length used to give the size of nuclei. It is equal to 10^{-15} meter. That is one million millionth of a millimeter. Nuclei are *small*.

[3] That is because there is no such thing. If you want to make a succession of observations you have to do it yourself.

"Ah, but there you're mistaken, Ma'am. The observations did have one thing in common. They all showed six electrons, even though they were at different positions each time. If the *atom* has six negatively charged electrons, then it follows that the nucleus must have six positive charges to make the overall atom neutral. This means it contains six protons. Each proton has one positive charge, so there are six protons all squashed together within this nucleus."

"If they are all positive and all very close to one another, how do they keep together? Surely they should fly apart." said the girl in some confusion. "I am certain I was told that positive charges will repel one another. Or negative charges for that matter," she added as an afterthought.

"Indeed they do," a new voice cried before any of her companions could answer. Dorothy turned to see a vision of EM rising tall within the electron amplitude. "The protons are so close together that they are thrust apart by forces millions of times stronger than those that operate over the wider confines of the atom. See for yourself." She pointed a long finger at the nucleus, just as the companions had earlier seen her point at a roadside pebble with dramatic results. Abruptly her motion stopped, and they saw that she was now swathed in colored bands that bound her so tightly she could do nothing.

"Don't even think about it, Sister!" said another voice that they had not heard before. "You know that your electricity is outclassed when particles feel my color forces. What color has joined, let no electric charge put asunder."

The vision of EM faded, still struggling (in a dignified manner) within the enveloping bands of color.

"The protons are glued together within the nucleus by strong nuclear forces. They are quite different from electricity, and their presence results in a sturdy and workmanlike product," continued the Inspector calmly in answer to Dorothy's original remark. He turned his gaze back to the nucleus. "I must now check on what else might be in there."

He dropped the nucleus into the palm of his hand and gently bobbed it up and down.

"This feels as though it weighs about twelve amu. That's atomic mass units, Ma'am. Twelve is about right. It means there must be six neutrons in the nucleus as well as the six protons. Neutrons weigh much the same as protons, though they do not have any overall electric charge. Because they are uncharged, they do not affect the number of electrons that the nucleus can hold. With my training, I can make quite a good guess at the weight of a nucleus, but I am required to check it more accurately."

He opened his canvas bag again and took out another shiny box, this one with a small platform on top. On this he placed the nucleus, flicked a switch and read off the weight from a digital display on the side.[4]

"What's this?" he said sharply, his easy and relaxed manner suddenly gone. "The weight is less than the sum of the protons and neutrons that compose it. We are being given short weight. This will never do!"

"If I might contradict you," interrupted the Lion, "it will *always* do. The mass of any compound nucleus will always be less than the sum of the masses of the protons and neutrons of which it is composed. The sum is less, not greater, than the parts because of the binding energy that holds the nucleus together. Any bound system such as a nucleus has a negative binding energy. This means that the components must be *given* energy before they can come apart, and without such energy input, they are captive. They are *bound*. It is all explained, at considerable length, in my book." The Lion produced from somewhere within the surrounding mists a book entitled *Particles Get in a Bind* by Isa C. Lion. He paused in the act of opening it. "But surely you must know about this," he said. "We will explain about mass and energy later," he added in an aside to Dorothy.

"Oh, of course I do," replied the Inspector, who had recovered his calm aspect. "It is all detailed in the standard product charts." From his invaluable canvas bag he produced a long, rolled-up chart, together with a handy folding display easel. He set this up and fastened the unrolled chart upon it. The chart showed a long curve that rose from zero at the left-hand end, climbed to a broad, gentle peak in the middle and then fell off gradually toward the right-hand side. Scattered along the curve were little drawings of nuclei of various types. At least, Dorothy *thought* they were drawings.

"This shows the binding energy per nucleon for various nuclei," he said, as though that explained everything.

"I beg your pardon," broke in the girl, who did *not* think it explained everything. Do you think you could explain that just a little?"

"Right, Ma'am. I know that these things are sometimes not too clear to people who are not in the trade. This little graph here shows the binding energies for nuclei of different sizes. The nuclei are arranged from left to right in the order of their electric charge—the number of electrons that each nucleus could hold in an atom. The lightest nucleus, hydrogen, is on the left,

---

[4] Measurements of nuclear size and mass are in reality a little more indirect than this. Size is found mostly from scattering experiments (like the alpha scattering mentioned in the text). Mass is determined by weighing large numbers of atoms (that is, ordinary matter) and allowing for the relatively tiny mass of the electrons, or more directly for individual particles with a device called a mass spectrograph. If you want to know how it is done, I refer you to the chapter on nuclear physics in your favorite college physics textbook.

R Gilmore

and the heaviest nucleus shown, uranium, is way over on the right-hand side. The diagram shows the binding energy for each nucleon. *Nucleon* is a blanket term for both protons and neutrons," he added as an aside. "As I was saying, it shows how much energy you would have to add to a nucleon to remove it from a given nucleus. This is the energy with which the nucleon is *bound* within the nucleus. You can see that there is no nuclear binding energy shown for hydrogen. Because a hydrogen nucleus is just a single proton, there is nothing for it to bind to."

"At the other extreme, the Inspector continued, "the neutrons and protons in the really heavy nuclei are not so strongly bound as they are in nuclei of medium size. This is largely because of the effect of the electrical forces you mentioned earlier. When a nucleus contains a lot of protons and every one is pushing *every other one* away from it, the binding energy is reduced. In fact, some of the heaviest are not quite as solid and well constructed as we might like. They are a little bit *unstable*."

NUCLEAR STABILITY

Atomic nuclei are composed essentially of protons and neutrons, which are known collectively as *nucleons*. They are held together by a short-range "strong nuclear force." This gives a sort of glue that sticks the nucleons together "where they touch." The protons also have an electric charge, and because they all have the same sort of charge, they repel one another. This tends to burst the nucleus apart, and there are in fact no stable nuclei that contain only protons (apart from hydrogen, and this is just one proton on its own).

The addition of neutrons serves to keep the protons further apart, and because neutrons are not identical to protons, the Pauli Exclusion Principle does not prevent neutrons and protons from going independently into the lowest energy levels available. The result is that most stable nuclei have roughly equal numbers of neutrons and protons. The ratio does vary a little, and there can be more than one type of nucleus for the same chemical element. These all have the same number of protons, however, so they have the same electric charge and capture the same number of electrons. They just have slightly different numbers of neutrons. Such nuclei are said to be *isotopes* of one another.

In the really big nuclei there are a lot more neutrons than protons. The electrical repulsion between protons is long-range, so *all* the protons in a nucleus repel all the others. The nuclear interaction affects only the neighboring nucleons, so the bigger the nucleus, the greater the relative effect of the proton repulsion and the larger the fraction of neutrons needed to separate them. Uranium-235 has 92 protons and 143 neutrons, for example, and even so, bits of this nucleus tend to fall off spontaneously (a process called α decay). Uranium decays very slowly, so there is still some of it around. More unstable nuclei all decayed long ago.

As he spoke, Dorothy noticed that one of the nuclei toward the right on the graph abruptly moved a couple of places to the left and a little higher up the curve. She commented on this behavior, which seemed rather a strange thing for a chart to do.

"Ah well, for reasons of economy we are using actual nuclei rather than drawings. This means that the unstable varieties do occasionally decay. You caught one in the act. It must have emitted an alpha particle. That's a couple of protons and a couple of neutrons bound together into one little packet," he added.

"Why doesn't it just throw out protons or neutrons one by one, if that is what a nucleus has inside it. Why should it go to the trouble of making them up into little packets?" asked Dorothy, quite reasonably she thought.

"It is because of energy again," answered the Inspector. "Whether something actually happens in the physical world is largely driven by an attempt on nature's part to release energy as kinetic energy, so that things can rush about all the more.[5] The protons and neutrons in the alpha particle are bound with a relatively large binding energy for such a light nucleus. Because they are bound, they have less energy than they would have if they were free and on their own, so this releases more energy for their movement. In particular, it means that if there is not quite enough energy available to eject a single nucleon, then it may be possible to emit the particles as a group that has an energy deficit of its own. This uses less energy than would have to be returned to each particle in order to make it completely free of all energy debt. It is rather as if you could not pay off the entire mortgage on your house, so you take out a smaller loan to cover part of it."

"I think I see that," replied Dorothy. She felt that she really did, although she was too young to have thought much about mortgages. "So in fact nuclei don't fling out protons or neutrons one by one. They emit them only in groups as alpha particles."

"That is not completely true, on several counts," replied the Inspector pedantically. His job as a supervisor of standards had tended to make him rather precise. "Nuclei with a *lot* of surplus energy will emit protons or neutrons, but any nucleus that has so much energy to spare will decay easily and quickly. Any nucleus that has lasted for a long time, which means any that you can find lying around, obviously decay only slowly and with difficulty. Such a nucleus will have very little surplus energy and can afford to emit only the particles within an α that still have some of their energy mortgaged."

"You are also wrong," the Inspector continued calmly, "in saying that only α particles may come out. Unstable radioactive substances may emit β and γ radiation as well as α.[6] The second is hardly surprising because a γ is just a photon of rather high energy. This is given off when excited states of a nucleus decay, in just the same way as a photon of lower energy is given off when the electrons in an atom change from one state to another. A previous decay process, such as the emission of an α particle, may well leave the residual nucleus in a disturbed condition. That nucleus may then shed energy by giving off photons, just as an excited atom will. Because of the much more compact size of the nucleus, the energies involved are much higher than for the electron states of an atom, though I prefer to say that the nucleus is more compact *because* the energies involved are greater."

---

[5] Nature's fondness for activity and for the conversion of other forms of energy to kinetic energy is a subject of the field of thermodynamics. It is discussed at some length in the first section of my book *Scrooge's Cryptic Carol*.

[6] Pioneers of nuclear physics found that radioactive materials emitted three distinct type of "rays" and called them, reasonably enough, α, β, and γ (alpha, beta, and gamma). They are in fact surprisingly different and are indeed primarily caused by three different interactions: strong, weak, and electromagnetic, respectively.

"The β particles are simply electrons," he added.

"Are there electrons *inside* the nucleus, then, as well as in a sort of cloud surrounding it?" asked Dorothy.

"No, of course not. Electrons are far too light to be compressed down to such a small size. Electrons are what we in the trade call *leptons.* This means, among other things, that they are not at all affected by the strong forces that squash the protons and neutrons together into such a small volume. The strongest force they feel is electromagnetism, and you have already been told that atom-size is as small as electrical forces are able to compress the electrons. That is why atoms are the size they are. A nucleus is *far* too small." He stopped speaking with an air of finality.

"Well, if there are no electrons inside a nucleus, then how can they come out as these β things?" A fair question, Dorothy thought.

"Wait and see," was the only answer. "Wait until you have seen the Wizard of Quarks and the Weak Witch. All, or at least something, will be revealed then." He closed his mouth firmly and would say no more on the topic, so Dorothy felt compelled to change the subject.

"Well then," she said. "If such a particle was thrown out when the nucleus decayed, why didn't I see it?"

"Because they are tiny and they move frightfully fast. They could be rushing all around you, and you wouldn't know that they were there. If you want to know when they are around, then you must call in the Detector."

"And who is the Detector?" asked Dorothy, though she felt, as she so often did recently, that she was about to be told whether she wanted to know or not.

"The best way to answer that is to introduce you," replied the Inspector as he took a small black instrument from his bag and pressed a few buttons on its face. There was a fairly long pause. Then they all saw in the distance a figure coming along one of the garden paths, beyond the limits of the magnified atom that surrounded them. They moved to meet him and were scarcely out of the expanded electron cloud when he stopped by their side.

He looked to be a rather scruffy person, with a thick moustache and a dingy cloth cap pulled down almost to his bushy eyebrows. He was wheeling a small handcart piled high with an amazing collection of completely unfamiliar equipment. Dorothy could see nothing there whose purpose she could begin to guess.

"Good morning," the Inspector greeted him. "May I introduce Dorothy

R Gilmore

and her friends? They want to know how one may detect the presence of high-energy particles. This is the Detector," he added rather unnecessarily, turning to the group of companions. "His calling is to detect where these tiny particles are, or rather where they were. They move so quickly that all you can see are the footprints they leave behind. Like a gamekeeper, the Detector sets traps in their path to catch them as they flash by. When they have passed through, their long-range electric interactions leave residual disruptions in his detectors, and he can find these afterwards. It may not be long afterwards by your standards, but on the time scale of the particles, they have long been gone."

"G'day," the Detector greeted them, making a half-hearted movement toward his cap in deference to Dorothy. "You want to know about detecting particles. That's tricky, that is. They're slippery, sneaky little varmints. They're so small and quick that there's no catching sight of them, 'cept for the ones with electric charge. When they move along, their charge gives them away right enough. They have their electric field, see. It spreads out far on all sides, however quietly they try to slip past. Great wings of electricity stretch far enough to be noticed by the likes of me. They jiggle electrons in atoms quite a ways from where the particles pass. When they have gone, they leave their surroundings full of light—photons emitted by excited electrons within the atoms—and also full of electric charge—positive ions and unattached electrons that have escaped their atoms entirely. There's a lot of litter about, and I pick it up with my instruments. Actually," he admitted, " 'full of light and electric charge' is a bit strong, but there's plenty left behind to show me sure enough that the little critters have passed that way.

"If they didn't have their electric charge, the particles could just shoot past and you would never know. It's the electric field that gives a particle away every time," he said with some satisfaction. "As a master Detector, I work through the electromagnetic interaction between these tiny particles and my devices. EM and I are like that," he added, making an improbable gesture with one hand that Dorothy took to indicate a close working relationship.

"Here you are then. Have a look at this," he continued as he turned and rummaged in his little cart. After a moment's searching he straightened and held up what looked like a hazy sheet of glass. "This here is a scintillator, this is. You can use it to show up particles as they try to creep past you, see." He walked across the garden path to a bed of uranium and held the scintillator sheet over it, shading it with one hand. In the depth of the material, they could see occasional little flashes of light as alpha particles from decaying uranium nuclei happened to pass through the device as they departed.

"There you are. Caught the skulking little critters. We have detected those alphas as they went past, so we know they were there. In the right situation, you can do more than just show that particles were there, a great deal more. You ought to see the work I do for the Kingdom of Cern."

"Where and what is that, may I ask?" asked Dorothy.

"It is a sort of kingdom, as you might expect. It's only a little way down the road from here," answered the Detector, who obviously knew it well. "It is quite small and mostly underground, but a lot of people from all countries visit there. They come to work on the particles. It is a great big collaboration of all races. If you want to know more about particles, I suggest that you go there. It's in that direction," he finished helpfully. "Just a bit farther down the Building Block Road. The road leads right through it, in fact, so you can't miss it. Off you go then. I might see you there and I might not," he added, shooing them off with what they felt to be unnecessary haste. They could take a hint, however, so all four took their leave of him, as well as of the Inspector and the Gardener, and they set off down the road again.

They traveled a little way along the road with no incident, and as they went the landscape gradually became rocky and then rugged and hilly. The path swung around the far side of a high outcrop of rock and vanished from their sight, while coming past the nearer side of the same hill, they saw a smooth highway that passed close to their track. Along this new road they could see a car approaching. It was a sports car—very low and smoothly curved with great chromium-plated exhaust pipes. All in all it looked built for speed, although at the moment it was traveling quite slowly.

"Why don't we ask if we can have a ride in this car?" asked Dorothy, who was beginning to feel a bit tired with all the uphill walking. No sooner had she spoken than the Lion leapt out into the middle of the road and reared up in the path of the advancing vehicle. The car slowed and swerved and seemed on the point of accelerating past by the Lion's side, but the Tin Geek grabbed hold of the back and lifted its wheels off the ground. The engine raced as the driver held his foot down on the pedal, but to no effect. Although the Geek's pipe-like arms looked rather spindly, his metallic strength was able to support the weight without difficulty.

"Oh, I am sorry!" exclaimed Dorothy to the driver as she came hurrying over. "I only planned to hold my hand out to ask you to stop." The driver grunted and gave no other answer. On closer examination, she saw that this was not so surprising. He was strikingly pink and hairy with floppy, triangular ears and a short snout. He was hunched over the wheel, which he gripped firmly with a dainty cleft hoof on either side.

"It would have done no good just *asking* him to stop. He is a road hog and as such does not have much consideration for other road users," said the Scarecrow as he came up to the car.

"Well, now that you have stopped, would you be so kind as to give us a ride? My feet are rather tired." The hog just grunted again. Dorothy took this to indicate consent, and they began piling into the car. The Scarecrow and the Tin Geek both climbed into the back. The Scarecrow, being basically a sack filled with straw, had no difficulty fitting into one of the tiny seats at the back. The Geek was not quite so flexible, but with his spindly metal legs folded so that his knees jutted out by his shoulders, he was able to squat there like a shiny metallic spider. Dorothy sat in the front seat by the

driver. When they looked at the Lion, however, it was clear that he was far too large to fit in by any means.

"Oh dear!" exclaimed Dorothy when she realized this. "What are we going to do? We cannot leave you behind."

"Don't worry about it," boomed the Lion, who for some reason did not himself seem to be at all worried. "Just go on without me. I shall soon meet up with you, in the Kingdom of Cern if not before." He turned away bravely, and at the same moment the Road Hog stamped down hard on the accelerator, and the car roared away. Dorothy glanced back at the massive figure of the Lion, now seeming rather sad and small as he shrank gradually from view.

Their driver continued to keep his foot, or rather his hoof, pressed hard down on the pedal, and initially the car surged forward smartly. As time went by, however, it did not seem to be moving as rapidly as might have been expected. Furthermore, Dorothy noted that the surrounding scenery had become strangely distorted.[7] When Dorothy commented on all this to the Scarecrow, who of course had been observing everything, he pointed at a road sign that they were just passing. It said $c = 40$. "What does that mean?" asked our heroine.

"It is telling us of a plan devised by the local town authorities to prevent excessive speeds on their roads. They have fixed the speed of light locally at 40 miles per hour, so nothing can go faster than this."

"Can a local authority actually do that?"

---

[7] I have not said in what way the scenery was distorted. It is not as simple as it is sometimes presented to be. A more detailed treatment, in a style similar to that of the present work, is given in the second section of my book *Scrooge's Cryptic Carol*, which is also available from Copernicus.

"Not normally, no. Here they have a particularly powerful legislative procedure, but even so it did not really have the effect they wanted."

As he spoke, they were approaching a pedestrian crosswalk, and just then a pedestrian decided to cross. Once embarked on her venture, she started across with an apparently suicidal lethargy. The Road Hog applied his full and considerable weight to the brake pedal, but it had no significant effect on the car's velocity. Although they were still traveling at well under the maximum of 40 miles per hour, the brakes scarcely seemed to affect their speed, and they shot over the crosswalk under the startled nose of the matronly lady who was attempting to cross.

"What happened there? Have the brakes failed?"

"No, it is not that," the Tin Geek answered. "It is simply an effect of traveling at a speed anywhere near that of light. You may note that people talk of *the* speed of the light. Its speed is just the same whether you are moving toward the light source or away from it. This may seem strange. Indeed, it is strange. It takes a special blending of space and time to bring it about. Anything that flashes past at such speed seems to be moving slowly. You saw this with the woman on the road crossing.

"Time and space seem to become askew when seen at such speeds. Motion is related to space and time, so all of kinematics has gone askew from what you expect. Energy and momentum assume a changed form. The speed of the car cannot rise above the speed of light. Indeed, it has great difficulty even in getting close to that value. However, the momentum can rise without limit. The application of a braking force reduces the momentum as you might expect, but unfortunately, this results in very little decrease in the velocity. As a result, the local authorities had to admit that their idea was a failure. They are now planning to build a pedestrian underpass."

The Tin Geek showed no sign of concluding his discourse. "You can if you wish still describe momentum as the product of mass and velocity. No object may have a velocity greater than that of light.[8] However, its mass appears to increase without limit as it approaches light speed. You find a similar behavior for energy. This also can increase without limit. Momentum gets larger both because the velocity is increasing and because of the larger effective weight. For energy, the increase is entirely due to the growth in mass.

---

[8] There is a possible confusion here. The speed c should really be called "the limiting velocity" or at least "the velocity of light in a vacuum." The velocity has nothing to do with light as such, and in fact light may travel at a lower speed in some cases. Light passes through your windows about 40 percent more slowly than the velocity c.

"Energy is equal to mass multiplied by the velocity of light squared.[9] Energy and mass are really the same thing. The old form for kinetic energy is found to be due only to the tiny increase in the mass that occurs even at low speeds. Even when an object is not moving, it has an enormous energy. This is called the rest-mass energy. You do not normally notice it because the mass does not change. You only see changes in energy."

ENERGY AND MASS

The theory of Special Relativity makes a lot of interesting statements about space and time. It also gives new relations among velocity, momentum, and energy. For our purposes, the important result is that mass and energy are really the same thing. When particles with high energy collide, some of this energy may be used to provide the mass of new particles.

The Geek paused after such an unusually long statement. He would have taken a deep breath, except that of course he did not breathe.

"Do you mean to say that our mass is actually increasing as we move faster?" asked Dorothy in astonishment.

"Yes, you might say that. Certainly it is true if you wish to say that momentum is simply mass times velocity. The mass measured at rest is called the rest mass. As the velocity approaches light speed, the increase in momentum derives mostly from the increase in mass. The increase in energy is effectively just this increase in mass. In fact, gravity affects energy, so it is simplest to say that the mass acted on by gravity is changing. Mass and energy are the same thing."

"Gosh!" said Dorothy. "If the mass or energy can change so much, doesn't it take a lot of fuel to get anywhere near the speed of light?"

"Certainly," answered the Tin Geek. "In fact, . . ." At that point the roar of the engine stopped abruptly. They all looked at the fuel gauge, which was registering "empty" in a definitive sort of way. The hog stamped his hoof

---

[9] Energy is given by the only physics equation to enter folk myth, namely $E = mc^2$. The velocity of light, c, is a constant, so energy and mass are in effect the same thing in different units. You can give a distance in miles or kilometers and convert one measure to the other if you multiply by the appropriate conversion factor. The constant factor c^2 is effectively just such a conversion factor.

down on the brake pedal, but as before, they did not slow down noticeably as a result.

On and on they went, with little observable effect apart from a smell of burning rubber. At last their speed did visibly decrease, and shortly after that they pulled to a halt at the roadside. There was something about the place that seemed vaguely familiar, and the feeling was greatly enhanced when around the side of a high outcrop of rock stepped the instantly recognizable figure of the Confident Lion.

"How did you get here?" asked Dorothy in amazement.

"I have been here all the time. This is where you left me," answered the Lion heartily. "You have ended up back where you started."

"Oh dear!" exclaimed Dorothy. "What a waste of time. At least it did not take very long. We were only away for a few minutes."

"On the contrary," corrected the Lion. "You were away for a good part of an hour. It's the time dilation effect, you know.[10] Don't worry, though, the time wasn't wasted. I managed to dash off the first three chapters of a new book."

"Do you mean we were going around in circles all that time we were away?"

"Well, yes. You were on a ring road, you know. Going around in circles in a ring is very much the tradition in these parts."

"Why is that? What parts do you mean? Where are we?"

"Oh, didn't you know?" asked the Lion innocently. "Just follow me, then."

They climbed out of the car, with various words of thanks to the Road Hog, who, lost in mournful contemplation of his empty fuel tank, merely grunted in reply. Together they followed the Lion around the obscuring rock and saw that the Building Block Road ran straight up to a massive door set into the mountainside. Over the top was a notice that stated, in very precise engraved letters, THE KINGDOM OF CERN.

---

[10] You might wonder why, because they saw the pedestrian on the crosswalk as moving so slowly, it turned out that when the group met up again, it was the stationary Lion who had so much time on his hands. Special relativity deals with steady uniform motion, and it says that such motion is entirely relative. Both a pedestrian and an occupant of the car will see time as passing more slowly for the other. So why is there any final difference? This "Twin Paradox" is treated more fully in my book *Scrooge's Cryptic Carol*. The critical consideration is that if you end up back where you started, you clearly cannot have been moving at a uniform speed in a straight line, as required by the conventional consideration of Special Relativity. There is an asymmetry in the motion.

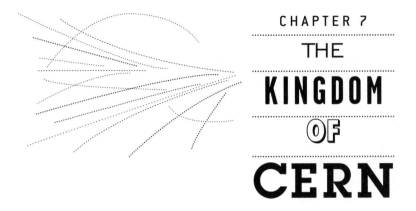

CHAPTER 7

THE
KINGDOM
OF
CERN

The group of travelers stood and stared in surprise at the massive door that blocked their way into the Kingdom of Cern.

"Did you know that we were this close to the entrance?" asked Dorothy.

"Oh, yes," responded the Lion with irritating complacency.

"And you just let us go off in that car and travel in a circle around a ring road when we could have stayed here," the girl continued with mounting annoyance.

"Yes, I did. But it *was* necessary, you know," the Lion hastened to explain, as he began to sense a certain antagonism. "You need to know that mass and energy are the same if you are to understand in any way what you will see within the tunnels of Cern. Besides," he added, "traveling around in circles in a ring is very much the fashion in Cern. You might as well get used to it. Now come along and let's not bicker." The Lion turned and stalked off with his tail proudly raised. The others followed him up to the doorway, where he rapped imperiously on a smooth area of rock beside it. The stone surface slid aside to reveal a hatch, behind which sat the guardian of the gate in a neatly pressed uniform.

"Let us in," demanded the Lion in a lordly fashion.

"Before I can do that, you must each correctly answer a question. Don't blame me. Everyone blames me, but it's in the rules. I just do what they tell me to."

The Lion seemed about to argue, so the Scarecrow gently pushed him aside and turned to the guardian. "We shall of course do whatever is required," he said diplomatically. "Do ask your questions."

"Right!" said the gatekeeper, looking at a printed list that he had before him. "Right, here we are. Now then," he continued, fixing the Scarecrow with a searching gaze, "tell me, what is the primary purpose of Cern?"

The Scarecrow thought for a moment, and then, being observant, he saw a small plaque mounted by the side of the entrance, near the engraved name sign. It read *Mission statement: Our primary purpose is to investigate the basic properties of matter.* He duly repeated this. "Correct," muttered the guardian, ticking it off on his sheet. "Now sir," he said, turning to the Tin Geek as they all clustered in front of his hatch, " would you please tell me the square root of 18769?"

The Geek immediately replied "137," and the guard ticked it off on his list. Next he turned to the Lion. "And you, Sir, would you be so good as to tell me the principal property that distinguishes fermions from bosons?" The Lion took a deep breath and settled down on his haunches. "In not more than eight words!" added the gatekeeper hastily.

"Fermions obey the Pauli Exclusion Principle," snapped the Lion, glaring at his questioner. "Bosons don't," he added, to use his full quota.

"And now, my dear," said the gatekeeper rather patronizingly as he turned to Dorothy. "Would you also like to enter Cern?"

"Yes please," she answered. Having come this far there was no other possible reply. "What is my question please?"

"That was it," returned the official, "and you have answered correctly as far as I can judge."

"That wasn't much of a question!" said the girl rather hotly. She felt somehow as though she had been cheated.

"No one said anything about the questions being difficult. It is the policy of the Cern authorities to encourage visitors, so I have to choose questions that people will be able to answer. You may now all come in," he finished as he pressed a button on the panel in front of him. With a low, grinding sound the massive door swung open to reveal a long tunnel leading deep underground.

Slightly fearful, the four companions walked into the darkness. As the daylight faded behind them, however, they found the tunnel was not so dark after all. There were lights at intervals along the walls and a stronger light up ahead. This increased as they came to the end of the entrance tunnel and walked out into a large, brightly lit cavern with people constantly passing through, entering and leaving by the many other tunnels that opened out on all sides.

"What a confusing place and what a lot of people!" exclaimed Dorothy.

"Yes, it is a very popular place. There are not so many permanent inhabitants, but there are a lot of visitors who come to work here. It is a collaboration between people of all countries and races, you know. Including some that you might consider slightly mythical," the Lion added as an afterthought.

"Their activities certainly look very purposeful. I wonder where we should go from here?"

"Perhaps we should ask at that little booth over there," responded the Scarecrow, who had observed in the center of the area a small stall that bore the label VISITOR INFORMATION. They all agreed that would be the best notion, and they made their way over.

In the front of the booth there was a hatch similar to the one beside the main gate, and behind the opening was a figure as neatly dressed as the other receptionist. He wore a trim, dark suit and a sober tie. He had smooth hair precisely parted and a small but exquisitely groomed moustache. He looked at them attentively.

"Excuse me," Dorothy addressed him. "Would you tell us please where we should go to find out about particles and such. We were told that this is the place we should come."

"It certainly is," he replied with what could only be described as precise enthusiasm. "There is a lot for you to see here, but the tunnels are too confusing for me to describe quickly. I am afraid that you would soon become lost. I had better guide you around myself. Excuse me while I back out of this stall."

The girl wondered why he would have to *back* out of the booth, and she glanced around behind it where she presumed the door was. The door was there, true enough, but she was surprised to see protruding from it the rear end of a horse. As she watched, this carefully backed out of the opening to reveal, in place of the neck of a normal horse, the torso of the neatly turned out man who had been addressing them. The front legs of the horse part were clad in trousers with precise creases. Dorothy was amused to notice that he wore spats overlapping his front hooves. The group all stared at him.

"Why are you all looking so surprised?" he asked. "Haven't you ever inquired from a Visitor Information Centaur before? Now come along. You have a lot to see."

He trotted off purposefully, and they dutifully followed. He led them along an instantly forgettable route through corridor after corridor, ever deeper under the Earth. Dorothy asked if they were descending into some sort of mine.

"In a sense," was the reply, "but only a mine of information. We tunnel not for gold and jewels but for something more valuable—for knowledge.

Here we seek to discover everything we can of the workings of the physical world."

As he was speaking, a group of seven humorously differing dwarves came marching out of a tunnel on one side and vanished down another on the opposite side. The onlookers caught a snatch of hearty song as they passed:

Hey ho, hey ho, it's after quarks we go.

The Centaur ignored this interruption and went on with his preliminary remarks.

"You had better start off with an introduction to the rings. That is what we call the circular accelerators that boost particles up to the high energies that we need. They are great rings of magnets that thread through tunnels in the rock around us.

"Why do you need particles of high energy?" asked Dorothy, rather afraid that she might miss a vital point.

"One of our main objectives is to search for new particles—particles, that is, that can exist, that exist in principle, and quite possibly have been important elsewhere in the Universe. It is just that at present they are not ready to hand *here*. Because they are so rare (indeed unknown) here, we have to make them. Particles usually have mass, so we must make that too. Mass is energy and conversely energy is mass, so if we have particles with energy to spare, then some of it can be used to make the mass of new particles."

"I told you that you would need to have heard about mass and energy being the same thing before you came here," put in the Lion, whispering with his usual complete audibility.

"When we have seen the accelerators," continued their guide, returning to his prepared speech, "we can go on to look at some particle detectors and observe the results they yield. But, first let's visit the main control room." He led them up to an imposing door and ushered them in.

"This is the principal control room for the whole system of accelerators, or rings, as we call them.. The operation of the machines is controlled from here." They saw around the room several control panels with attached working surfaces. Papers and notebooks cluttered these surfaces in a haphazard but purposeful way. At a keyboard in front of the nearest panel sat a squat figure wearing a T-shirt with an unusually loose neck, to accommodate the massive bull's head that rose from his shoulders. On the back of his shirt was printed the message *My folks went to Crete and all I got was this dumb T-shirt.* He was peering intently into a sort of computer screen in front of him and making occasional adjustments on his keyboard.

"This is the duty engineer, who keeps a continual check on the operation of the machines. As you see, he is able to monitor every component on his special Minotaur Screen. The purpose of the machines is to accelerate beams of particles to high energies. This is done by accelerating them repeatedly as they travel again and again around a great ring of magnets. Here is the Keeper of the Beam, who will explain how this is done."

At this point a stocky figure entered the control room. He had glossy black hair and a bold black moustache. He wore a safari shirt and corduroy breeches. These latter were held up by a thick leather belt to which was clipped a long, coiled whip. The whip was peculiar in that it hissed and crackled, rather like the vivid discharges of EM's electric staff.

"You want to know how we train particles and make them bend to our will, do you?" he asked rather aggressively. "Well, you have come to the right person. I am in effect a *particle tamer*, rather like the Lion tamers you may have met in your world."

The Keeper and the Lion locked gazes briefly, but they both decided not to pursue the point.

"I soon show the particles who is boss," continued the Keeper, unclipping his remarkable whip from his belt. "Any that carry an electric charge must obey my electromagnetic whip. Let me show you. I can see a few protons skulking over there in the corner. They will make a good demonstration for they are the easiest to train."

He approached the group of particles and lashed his whip out toward them. The electric field acted on the positive proton charge, accelerating the particles along that direction. He now changed his tactics and coiled his whip, then held it above the fleeing protons. Of one accord they veered to one side in their flight.

"There you see two uses of electrical forces in the control of particles," he commented as he went about his work. "The direct electric field will make them move sharply at my command. The magnetic field produced by a circling electric current will then control the direction of their movement. I can make them jump through hoops, or at least I can readily make them circle around rings at my command." To illustrate, with successive little flicks of his whip he made them move ever faster, while he guided them around so that they ran about in a circle.

ELECTROMAGNETISM

The name is a composite of *electricity* and *magnetism*, which turn out to be the same thing.

Electrical forces act between electric charges, and there are two varieties of these with different *signs*: positive and negative. Charges of the same sign repel each other, whereas opposite charges attract. Around the charges there is an *electric field*, and this will accelerate the charges along the direction of the line that joins them.

If the charges are moving, they constitute an *electric current*. An electric current, whether in a coil of wire or within iron atoms, will generate a *magnetic field*. A magnetic field will exert a force on a *moving* charge that acts at right angles to its direction of motion, so that the particle's path will curve around in a circle.

"You can see for yourselves the overall behavior of the accelerator on that diagram up there." He pointed to a large display mounted high up on the wall. It showed three rings of increasing size, linked together by tangential lines. A small cluster of lights moved over the surface of the board. The lights had just entered the smallest ring and proceeded to race around it many times.

"Those lights represent a burst of protons that has been injected into the smallest accelerator ring and is now circling around it. The protons are forced to travel in a circular path by large electromagnets spaced around the ring. These magnets deflect the protons and bend their path to fit the ring. Each time a bunch of particles orbits around, the particles pass through a cavity in which an electric field acts on them. This field gives each of them a push and increases its momentum slightly. They are soon moving so fast in their hectic flight that they travel almost at the speed of light. After that their actual speed doesn't vary much, but their momentum still increases. As the particles gain momentum, the machine increases the strength of the magnetic fields accordingly so that the particles keep moving precisely in the same path, along the center of a narrow beam pipe.

"What is in the pipe?" Dorothy felt she should say something to show that she was suitably attentive.

"There is nothing in it. That it is to say, it contains as close to nothing as we can arrange. The particles travel in a good vacuum. Otherwise, they would be scattered from collisions with atoms in the air and would consequently be lost from the beam. When they have made many orbits of the ring and have been accelerated each time that they go around it, they will have been given a large momentum. The strength of the magnetic field that is necessary to keep them in the same path, when they have such an increased momentum, has become as great as it is practical to produce. At this stage the particles must be transferred to another, larger ring."

The cluster of lights on the wall display, having circled the smallest ring many times, abruptly switched to the line that connected the smallest to the intermediate ring. They quickly transferred to this ring and began to circle around it in turn.

"In this larger ring the course of the particles is not so sharply curved, and so a weaker magnetic field is able to keep them in their required orbit. They may thus be accelerated to a yet higher momentum and energy before reaching the limit of the magnets. When the beam of particles has been given as high a momentum as this ring is able to contain, they are then transferred to the final ring. This has a much greater diameter and a more gently curved path and so can handle an even higher particle momentum.

"When the particles have been accelerated to their maximum energy, they are deflected and focused so that they collide with other particles—

either with the nuclei of atoms in a stationary target or with a beam of particles that are circling around the machine in the opposite direction. This latter option is implemented in a *colliding beam* accelerator, in which the momenta of the oppositely moving particles cancel to zero. All the energy in a collision can then be available to produce the masses of new particles."

On the diagram the lights moved around and around the largest ring, and then, abruptly, they simply vanished. This was because the consequent collisions were not part of the acceleration process and were no longer the concern of the Keeper of the Beam. He had delivered the beam, and thus had accomplished his task, but it was nonetheless an anticlimax. A new burst of particles began its journey, and a cluster of lights again entered the smallest ring.

"It seems that everything you do is using electrical or magnetic forces. I thought you were concerned here primarily with this strong nuclear interaction I have been told about," said the girl.

"Oh, indeed. Our *purpose* here is to examine the strong forces. Our problem, though, is that the range over which these forces extend is exceedingly small—barely more than the size of a nucleus. We can sense no effect over distances large enough to perceive in the day-to-day world, and the only way to probe interactions at such short distances is to make particles actually collide with one another and then examine the results. To do this we must be able to accelerate particles, guide them to a collision, and detect them afterward. As I told you initially, the accelerators, the detectors—all of the machines we use—have to deal with the particles through their electric charges. This is really the only way to get a handle on them over any reasonable distance. Their strong interactions are of much too short a range to be of any service in this.

Although our objectives are in the nuclear domain, our methods, like most things in your world, are within the sphere of EM. The devices that serve us are all hers, and she is always with us. See!" he added, gesturing at a large window that looked out over the control room. Beyond it through a blind, they saw dimly, the tall figure of EM standing before a desk. Behind the desk was a massive figure shaped of shadow and suggestive of power—a mighty power beneath the Earth.

"Who is that?" they asked their escort, who had been standing silently to one side throughout the Keeper's demonstration.

"That is the Director of our whole accelerator division. He is in overall charge of the acceleration of all the particles in our accelerators. He is the Ring Master for our particle circus. He is the Lord of the Rings," answered the Centaur in awed tones. "But come now," he continued. "Let us move on

to the experimental hall, the place where the collisions happen and the particles reveal their secrets." He led them out of the control room and down a corridor. They passed a room with its door ajar, and inside they saw a group of small men clad in green jackets and knee britches, all clustered around a completely circular rainbow. In the background they could hear a sweet voice singing *"Somewhere over the rainbow."* The guide noticed the direction of his charges' glances and explained the scene.

"That is part of our design effort for future accelerators. The rainbow you see there is an example of what we call 'blue sky' design by our team of LEPrechauns.[1] The design has one serious flaw."

"And what is that?" three voices asked as one.

"Simply that when you close a rainbow into the ring shape of an accelerator, it has no end, and they will need to find the pot of gold at the end of the rainbow before they can afford to build the accelerator. Come now," he continued before anyone could react to this last remark. "The main machine has just been shut down temporarily, so it might interest you to approach the experimental hall through the actual accelerator tunnel."

As they walked on down the corridor they passed a number of men in dark suits. "Those are Gnomes," confided the Centaur.

[1] LEP is the acronym for the Large Electron Positron collider at the CERN laboratory in Geneva.

Dorothy looked around after them. They still looked just like men in dark suits. "They do not look like my idea of gnomes. In fact, I don't think they look in any way mythical," she declared.

"Oh, they're not. Far from it. They are bankers from Zurich. They are helping us with our finances. Although our primary objective is knowledge, pure knowledge, finance is still very important to us."

The Centaur then led them through a massive door that was standing open, and soon they were in a long, curving gallery. It was possible to see that it curved only if you looked far into the shrinking perspective in the distance. At first glance it appeared to be quite straight, so gentle was the curvature. To one side, spaced along the tunnel, were a number of massive shapes, like strange metallic sepulchers. They were painted in bright colors, and each was threaded by a metal pipe not unlike a domestic water pipe apart from its fine finish. These were the magnets and the beam pipe that ran through every one of them. Within their metal casings were the coils of thick metal cables to produce the intense magnetic fields required. Between the magnets an incredible array of equipment clustered around the pipe. There were the pumps that created the vacuum; there were sensors of all sorts to confirm the position of the particle beam; there were devices whose purpose it was hard to imagine. In all probability, there were devices whose purpose no one could now remember. Around, below, and through all of this there were great bundles of cables and feed pipes painted in coded colors to show their purpose. The group walked through in silence. No comment seemed appropriate to this complexity.

Finally, after walking for what seemed like miles beside this diverse but repetitive machine, their guide led them from the tunnel. They passed out through another heavy door and into a sort of assembly area, a cavernous room with a high ceiling. Massive pieces of equipment were scattered around. For no particular reason, a stout post nearby caught Dorothy's eye. It had a platform on top, and when she peered up she could see a gargoyle-like creature with large eyes perched on it.

"Why is he there?" she asked.

"He is one of our night watchmen. He is a ghoul, and they have very good night vision. He is on top of a post," he continued, driven by a long-standing habit of explaining things to tour groups, "because that gives him the best field of view over the whole area."

"And why is the post mounted on little wheels?" Dorothy had just noticed this feature.

"Ah, that. That is just for convenience. The management keeps moving the ghoul posts, you see."

The Centaur trotted elegantly into the middle of the floor and pirouetted around to face his audience, who had followed obediently behind.

"This is an experimental hall where the particle detectors are assembled before they are placed in the beam from the accelerator. I see that a specialist in this subject has just arrived. I shall ask him whether he will come over and talk to you."

They looked across the hall and saw approaching the Detector, whom they had last seen in the Atomic Garden. He greeted them affably.

"I see that you do not have your handcart of equipment his time," remarked the Scarecrow, who was good at observing this sort of thing.

"No, mate," chuckled the Detector. "This job is a bit too big for that little dingus." He reached into his pocket and produced a small whistle, which he blew with a piercing blast. A wide door at one side of the hall opened to reveal a broad tunnel beyond and, waiting in front of the door, a fleet of enormous trucks. These drove in and parked around the edge of the area. "This stuff is mighty heavy, some of it. We need a bit of heavy lifting here to get it into the right position."

He ambled over to a plinth standing in an alcove at one side. On this stood an antique brass oil lamp that he proceeded to rub vigorously. Nothing happened for a moment, and then a voice said, "Go away. It's our tea break." Everyone stood around looking slightly uncomfortable for a while, and during this time, Dorothy noted that there were letters finely engraved on the curved brass side of the lamp. They read

AUGAD
Wishes at union rates only

"What does that stand for?" she asked.

"It means it is the headquarters of the *Amalgamated Union of Genii, Aftreets, and Djinn*," replied their escort.

This exchange had taken a while to complete, so the Detector thought it worth giving the lamp another sharp rub. This time there was an immediate effect. A cloud of smoke spurted from the lamp, thickening and billowing up toward the roof of the great cavern. In the air over their heads it thickened and coalesced into the monstrous torso of a figure with arms folded over a boiler suit, the lower part of which faded away into a residual wisp of smoke. He scowled at them from beneath a cloth cap. A second cloud of smoke and a third followed from the lamp, and soon there were three gigantic figures hovering above them. The central apparition did not exactly bow, but it at least gave a slight nod. It spoke.

"What is your wish, my *master*?" The last word carried a distinctly ironic emphasis. "Your wish is my command. Though if we don't like it, it will be referred to the next meeting of the brotherhood," he added darkly.

"Stop messing about and just shift this lot," retorted the Detector. "I need you Djinn to have everything in position and ready to go before the accelerator beam comes on again. So get on with it," he finished ungraciously.

"How will they be able to lift such large objects?" the girl asked the Centaur, who was standing quietly nearby.

"They will use a Djinn Sling. Just watch."

Despite their earlier truculent attitude, the three forms moved about smartly and in unison. Although they *looked* a bit cloudy, they did not seem to have any difficulty in carrying great tanks and blocks of iron, each suspended from a complex cradle of thick cables. They soon had the trucks unloaded and promptly assembled the loads into a large and compactly massive structure, completely sheathed in an iron magnet, pierced here and there by holes through which sprouted thousands of cables laid in thick bundles. The device was positioned near a concrete wall that apparently divided this area from the accelerator region.

"There, the finished detector!" said the Detector proudly.

At that moment bells suddenly began to ring everywhere, a horn sounded urgently, and lights flashed over warning notices around the area. "It's the accelerator!" cried the Centaur, rearing up in alarm. His neatly attired human half was lifted well above Dorothy's head, and she had an inappropriate vision of him as a horse trying to rid itself of a tailor's dummy with which it had become entangled.

This was not the time for such fancies, however. The three Genii worked ever more swiftly under the direction of the Detector. They quickly shifted the huge concrete blocks that shielded the access to the accelerator, moved the apparatus into the area beyond, and hastily closed off the opening again. All the while, the warning signals continued.

R Gilmore

The accelerator was starting up now, gathering itself to hurl bursts of electrically charged particles through this new detector at energies up to many times their rest masses. The noises stopped, the lights settled down to a steady warning glow, and beams of particles flashed in opposite directions through the central chamber of the apparatus.

Everyone clustered in front of a large screen that, remarkably, showed the inside of the complex equipment, giving a clear view of the central chamber, past all the containing layers of magnetized iron and other material.[2] The detector chamber was filled with thousands of fine wires, each of them the center of an intense local electric field. In this field the ionization from the particles gave brief electrical discharges, so the particles left fiery footsteps as they passed, their tracks clearly visible on the screen in front of Dorothy.

Most of the beam particles simply rushed straight through without a collision, but occasionally two collided head-on. When they did, they sprayed out to either side, and the watchers could see that although only two particles had collided, many more than two came out of the collision. Dorothy wondered aloud why this should be so, and this time it was the Lion who replied. Dorothy was a little surprised that he had managed to be silent for so long.

"You are seeing *particle production*. You remember how it was explained during your car ride that energy and mass are essentially the same thing? Well, in this case, the colliding particles both have energies much greater than their rest masses. When they are moving along on their own, they cannot part with any of this energy, because it is goes with the momentum they have. When they collide head-on, their momenta cancel. They splash together to give a composite object. This has very little momentum, but it still has all the energy that was there before. Because the energy is no longer needed to partner momentum, it is available to make new particles, and that is what has happened."

Now that he had the floor, the Lion was not about to relinquish it. "The particles produced are usually bosons, because bosons are not conserved. If you have the energy available to make their mass, then, all things being equal, you can create bosons freely. They may be created singly or several at a time; it depends on the energy you have."

"What do you mean by 'all things being equal'?" asked Dorothy. She suspected that this proviso hid unknown complications.

"It depends on what other properties the bosons have. They may carry an electric charge, for example. Electric charge is conserved in its own right.

---

[2] This is pretty much what really happens. The view you see is constructed by computers that use the information carried by all the cables.

PARTICLE COLLISIONS

Elementary particles are much smaller than the wavelength of visible light and so cannot be seen by the unaided eye. You have to examine them with something that has a smaller wavelength, and, as given by the de Broglie relation in Chapter 3, this means particles of high momentum.

Beams of particles can be made to scatter from one another, and examination of the particles that come out of these interactions reveals much about the particles. A detailed analysis shows, for example, that the scatter is from point-like objects within the beam particle: These are the quarks.

The accompanying picture shows a great shower of secondary particles created from the energy of two interacting beam particles that enter from the left and right and collide in the center. This picture is reproduced with permission from the CERN laboratory in Geneva.

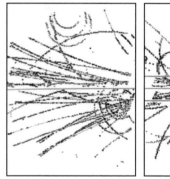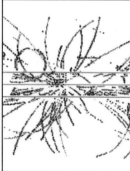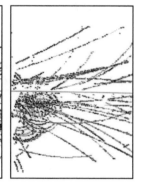

You cannot just create a positive or a negative electric charge. You can, however, create a positive *and* a negative charge together, because then the total change is zero and you haven't actually created any charge at all. Many of the particles that came off would have been the type known as pions, for these are very light and do not require too much of the energy to create their mass. With pions you get more particles for your money, or rather for your energy, but there might well be few heavier particles also.

"You *can* get fermions as well as bosons," he continued. "But the fermions must come in pairs. Fermions themselves are conserved, so you must balance any fermions produced, just as you must balance any electric charge produced. Each fermion must be accompanied by an appropriate antiparticle."

"What might an antiparticle be?" asked Dorothy patiently.

"Every particle has one. An antiparticle is in a way like the particle itself, but it is everything that its companion is not. If a particle has positive charge, then its antiparticle is negative. If a particle has negative strangeness, then the strangeness of its antiparticle is positive."

"And what is strangeness?" asked Dorothy, but the Lion swept on, apparently without hearing.

"Particle and antiparticle are twins, but not identical twins. They have exactly the same mass, for example, but otherwise each provides what the other does not, and they can cancel out each other's properties. They may be created together if enough energy is available, just as a boson might be."

During this discourse, other collisions between incoming particles had been shown on the screen. It was noticeable that the paths of some of the outgoing particles were distinctly curved, some much more obviously so than others.

"That's the magnetic field, that is," proclaimed the Detector proudly, if rather cryptically. "I don't put all that iron around my detector for nothing, you know. It's a magnet and it gives a strong magnetic field in the middle of the whole caboodle. Any charged particles will bend in the field, they will. Each one follows a curved path. The lower its momentum, the more curved its track. That one . . . ," he added, pointing to a glowing line on the display that had curved right back on itself. "That one don't have much in the way of momentum at all. The ones that have a *lot* of momentum look as if they are going straight, though in fact they don't quite. It is just that when they have really high momentum, the magnetic field isn't able to force much curvature on them."

Just at that moment something dramatic happened in the detector. On the screen a great shower of particles burst out, concentrated in several distinct jets. It happened so quickly that no one could be quite sure, but it seemed that each jet of particles burst forth from a slightly different point.

"Whatever was that?" asked Dorothy.

"That was a rare event." This time she was answered by the Centaur, who was used to explaining such things to visiting groups. "You saw, fleetingly, the production and decay of an unusual and massive particle. It split into a number of others, and then these in turn reverted to a bunch of more familiar particles: pions and the like. Something new and wonderful came into the world, stayed but for a moment, and then faded again to the familiar and mundane."

"It was so quick that I couldn't really see what happened. What a pity we couldn't keep it a little longer!" lamented the girl. "If only there were some way to preserve it.[3] Is it likely to happen again soon?"

"You cannot really say. Such occurrences are quite random and unpredictable. Sometimes it is a long time before it happens again; sometimes another such event occurs almost immediately."

Scarcely had the Centaur spoken than another dramatic burst of particles erupted across the screen. This time, however, the image did not fade. Instead, the central portion of the picture *peeled off* the screen. It came fluttering out like a sheet of paper and floated over their heads. They all turned to follow its progress and saw that it was heading toward a young woman who had come up behind the group. She was wearing stern secretarial glasses and sported an unusual hairstyle. More accurately, she sported unusual *hair*. Sprouting from her head was a collection of thin but active snakes, weaving

---

[3] In actual practice, computers should record all the events displayed throughout an experiment, so there is no question of one fading unnoticed, as here.

and hissing at one another as they twisted around. Even before anyone spoke, the Centaur could see that an introduction was needed.

"Meet Zola. She is a Paper Gorgon and works on publications. You know how the gaze of Medusa, the original Gorgon, turned everything she looked at to stone? In the case of Zola here, whatever she observes is converted to paper and subsequently published in the technical literature. Be careful that you do not meet her eye, or you may find yourself nothing more than a character on a printed page."

The Gorgon went over to a little table nearby, on which stood several typewriters. She sat down, and her unruly snakes promptly disentangled themselves and blended into a pattern of smooth co-operation as each one poised over the keyboard of one or another typewriter and began to type with incredible speed. Pages of description leapt from the rollers of the machines and joined with the illustrations that were still drifting toward her from the detector to form complete publications. These went fluttering off across the cave and into a side tunnel.

"I must see where they are going!" exclaimed Dorothy, and she ran down the tunnel after them. Perforce the others followed.

It was like being in some sort of snowstorm. The tunnel was filled with a regular blizzard of papers, so numerous were the publications unleashed upon the world. Immersed in the middle of all this was a frantic figure wearing a track suit and carrying an enormous butterfly net. He was leaping and capering among the publications, sweeping his net to and fro through the crowded air. In many cases, the mesh of his net passed right through the pages, and nothing at all was left behind. Sometimes, however, the net captured something from the writings. A nugget of valuable information remained in his net, trawled from the flurry of paper, Whenever this happened the figure extracted the treasure with great satisfaction and placed it in a leather satchel hanging at his side.

Abruptly, alerted by the sound of their approach, he stopped his reckless bounding and turned to meet them.

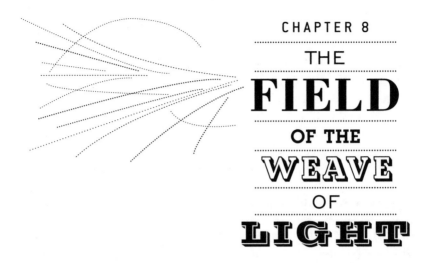

CHAPTER 8

THE
FIELD
OF THE
WEAVE
OF
LIGHT

The group of companions walked up to this stranger as he stood watching their approach and clutching his butterfly net almost defiantly. Now that they could see him clearly, he was revealed as a small, elderly man with a bald head and side-whiskers. He was distinctly too stout for the strenuous activity they had witnessed.

"Hello," Dorothy accosted him in a friendly fashion. "We couldn't help wondering what you were doing."

"Ahem, well," he began uncertainly. "I am collecting. That's it," he continued more confidently. "Collecting particles, that is. I am a bit of a collector you know, a hadronologist."

"Can you tell me what a hadronologist is?" Dorothy whispered in an aside to the Lion. She was pretty sure that he would have an answer. "Someone who collects hadrons," was the Lion's surprisingly brief response. "What is a hadron then?" whispered the girl impatiently. "It is a particle that experiences the strong interaction. Hadrons are affected by the strong forces, whereas leptons are totally indifferent to them." "Oh!"

HADRONS AND LEPTONS

Elementary particles are divided into hadrons and leptons. The hadrons are affected by the strong interaction. The leptons are not.

That is essentially all there is to say.

Their new acquaintance had been waiting patently during this whispered exchange. When it had finished, he held out his satchel. On closer examination they could see that it was a sort of portable cabinet with flat sliding drawers.

"Would you care to see my collection of hadrons?" he asked rather shyly. And without waiting for an answer, he began to open drawer after drawer. Each drawer was very shallow, and each displayed an array of particles in one or other of two different patterns. One option that occurred frequently was a hexagonal array with two particles at the center, giving a total of eight particles. The alternative arrangement, which was less well represented, was a triangular array with four particles across the top row and one less in each of the rows below, making ten particles in all.

"How do you like my octets and decuplets?" asked their new acquaintance. "Particles fit naturally into these classifications when you sort them out in a sequence in terms of electric charge and strangeness. The different trays hold particles of different spin, and of course some are baryons and some are mesons."

"What's strangeness?" asked Dorothy, but again she did not get an answer, as the collector went on talking about his hobby. She had decided that she wouldn't even bother asking about baryons and mesons, but she quickly discovered that she didn't have to.

"Some hadrons are bosons. These are called mesons and they arrange themselves into octet groups. Other hadrons are fermions and they are called baryons. They can be found in either octets or decuplets. One of the baryon octets, this one here. . . ," he added as he opened a drawer to show yet another group of eight particles. "This contains the earliest members of my collection. They are the proton and the neutron. The proton in particular has been known for a long time now, because on its own it forms the nucleus of hydrogen. This one is still as fresh as the day I began my collection, many years ago. It may decay eventually, but not for another 10^{33} years or so."

"How long is that?" the girl asked curiously.

""How long is 10^{33} years? Well, if you consider how long one second is in comparison to the lifetime of the Universe, then the lifetime of the Universe is to . . . Umm. . . . Anyway, it's a very long time. It should see me out."[1]

"Do you collect only hadrons?" asked Dorothy to change the subject.

"Oh yes! What else is there after all? There are one or two leptons, I

---

[1] What the hadronologist is trying to say is that 10^{33} years bears about the same relation to the lifetime of the Universe as this latter does to one second. Can you visualize that? Neither can I. As the man says, it's a very long time.

grant you. Leptons are particles that do *not* feel the strong interaction," he added by way of explanation, "but there are only a few of them. There is the photon, of course, but there is only one of it. No, it has to be hadrons. There are hundreds of them. You ought to see the collection in the Emerald (and Ruby and Sapphire) City."

"That is something we should like to ask you," she cut in quickly now that the subject had come up. "We are trying to get there, as it happens. How should we go from here?"

"Just follow this corridor and it will lead you to a wide tunnel that leads in turn to a gate out of the Kingdom of Cern. The Building Block Road runs out of this gate, and if you follow that, it will take you right to the city."

They thanked him and hurried on their way. The route was just as he had said. Soon they were outside the linked caverns of the Kingdom of Cern and once again walking along the Building Block Road under an open sky. The country here was quite flat, and in the distance they could actually see the walls and towers of the city.

"I wonder why it is called the Emerald (and Ruby and Sapphire) City?" wondered Dorothy aloud. "It looks to be almost completely colorless as far as I can see."

Before anyone attempted to respond to that conversational opening, they came to a place where the Building Block Road entered a wide meadow. It continued across this as a mere track in the grass, and by its side they saw a sign mounted on a post. It read:

> You are now entering an
> **electromagnetic field.**
> Severe danger of
> **virtual photons.**
> You have been warned!
> signed **EM**

"Photons! Didn't someone tell me that photons are just light. There's nothing to worry about, then. I do not see how light can possibly harm us. Come on, let's get on to the city."

"Well now, I'm not so sure . . . ," began the Lion, rather atypically. But by then the others had already started along the track, and he rushed to follow them.

As she walked across the meadow, our heroine looked down at the grass beneath her feet and noticed that a sort of flickering mesh covered it. She observed this just as she was about to take a step forward, and as she at-

tempted to lift her foot, it stuck to the ground, firmly and quite unexpect-
edly. Caught off balance, she tripped and fell, and as her arms touched the
ground, they too were captured. Dorothy found herself sprawled on the
ground and unable to move her limbs by even a fraction of an inch. It was as
though they were held firmly by bonds of steel.

She was just able to look to one side or the other, and she could see that
her companions were in the same predicament. The Tin Geek was spread
out as though he had been welded to the ground. The Scarecrow looked as
though his straw-filled limbs had been stitched to the meadow. As they were
falling the Lion exclaimed, "Virtual photons, I should have . . . " and then
was silent. This seemed even more remarkable than her present awkward situation, and Dorothy struggled to turn her head toward him. When she managed to look across, she could see why he had stopped. Not only were the
Lion's limbs trapped, but his chin was also resting on the glowing mesh and
was held so firmly so that he was unable to utter a word. Beside his hand,
where he had been holding it before he was brought low, was a book. It bore
the title *Our Bonds Are Light* by Isa C. Lion.

The girl wondered what she should do. Actually, there did not seem to be
much point in wondering this, because she could not *do* anything. She looked

across the meadow and saw, some way off but approaching rapidly, a tall, agile figure. She recognized the Adjusting Acrobat, whom she had previously met shortly after she encountered the Tin Geek. He was now making his way, apparently without difficulty, across the treacherous field. His progress was unusual. Sometimes he moved with a sort of skipping motion, sometimes he turned somersaults or cartwheels, and sometimes he leapt high in the air. His toes twinkled as he slipped past the net of photons that had captured Dorothy and her friends. He did seem to be remarkably light on his feet, although the point was that there *wasn't* light on his feet—he managed to avoid all the photons that had tripped and trapped the others. However he did it, he kept going without at any time being captured, and he soon reached them.

"Hello," he remarked brightly. "You do not look very comfortable. Would you like me to help you up?"

Three of the companions immediately gave their enthusiastic assent. The Lion was not able to do this, because he could not speak, but he did thump his tail once on the ground. He had probably intended to thump more than once, but the first time his tail landed, it too was caught fast.

Somehow, and with surprisingly little difficulty, the Acrobat released them and helped them to their feet.

"How do you manage to do that?" they all asked him.

"It's just a matter of being well adjusted. I am the Adjusting Acrobat, and I can usually adjust to any circumstances. You, however, have no way to escape electromagnetic bonding, the bonding produced by your interaction with virtual photons in this field of light. It might be better if you were to Keep Off The Grass." As he was speaking he had led them, without anyone really noticing (except perhaps the Scarecrow), to an area free of the Weave of Light. There a few tents and stalls were scattered about in random fashion.

"This is the Virtual Fairground and Photon Exchange. It serves as an exposition of virtual particle emission and exchange. A photon carnival, where we celebrate the role played by virtual particles in your reality."

"What are virtual particles?" asked Dorothy. She half expected to be ignored as she had been recently, but the Acrobat replied, though rather cryptically.

"They are particles that live beyond their means." As he spoke, he began to juggle a number of objects. The girl did not quite see where they had come from, but she saw him toss three objects from one hand to the other— except that she now saw there were actually four of them. No, there were actually six of them, no three, no two, no four. The number of items he juggled was forever changing as she watched.

"Are you a conjuror?" she asked him. It seemed to be the only possible explanation.

"No, this is no trick. These are bosons and their number *is* changing. I do not have the energy to create their mass. This is the rest mass. Most particles possess this intrinsic personal mass even when they are not moving, and it is one of the features that makes them what they are. Because I do not have the energy, I cannot create *real* particles as you saw happen at Cern. Particles can, however, *steal* their own existence. We spoke before of the Uncertainly Principle that relates momentum and energy. I told you how squashing a particle down into a small volume gives it momentum, for you cannot have a spread of momentum unless you *have* momentum. The same thing applies to time and energy. Localization in time results in a spread of energy, energy that would not be there otherwise. Particles can steal energy from the Universe, but they can't keep it very long."

VIRTUAL FAIRGROUND

R Gilmore

VIRTUAL PARTICLES AND THE TIME–ENERGY RELATION

Just as there is an Uncertainty Principle for energy and momentum, so too is there one for spread in time, Δt, and spread in energy, ΔE.

$$\Delta E \Delta t = \frac{\hbar}{2}$$

As before, the uncertainty may be greater than this, because you can always be *more* uncertain about something. This equation gives the relation between the irreducible *spread* of time and that of energy. In the short term, energy is not well defined. For a sufficiently short time, the indefiniteness in the energy may be enough to allow the existence of the rest mass of a particle. Such particles may literally "come from nothing," though only for a brief life. They are called virtual, but they are an essential part of our reality and provide all the interactions that hold the world together.

"Do not be confused by the names *real* and *virtual* for particles,"[2] the Adjusting Acrobat added sternly. "Virtual particles are every bit as important a part of your physical reality as are so-called real particles. It is just a matter of their energy account.

"I am sure that you have been told already that energy is conserved: that energy may change from one form to another but that the total amount of energy never changes. This is true enough in the long term. You cannot just create energy. The laws of classical physics say that you cannot have more energy than was there before. On the quantum scale, however, the scale of $\hbar$, you can manage to steal a little energy *for a short time*, though you soon have to put it back. The more energy you take, the sooner it must be replaced. During such short times, stolen energy may be used to make matter: bosons or even particle–antiparticle pairs. Because most particles have a mass, and because mass is energy, these particles do not just steal energy. They steal their own existence. These are *virtual particles* that exist by the grace of the 'Uncertainty Principle.' They live in whole or in part on stolen energy—on energy stolen from the Universe itself."

"Gosh!" said Dorothy. "But if their existence is so precarious and must end when their energy vanishes, surely such particles can have little effect on the scheme of things."

"Wrong!" cried the Acrobat, loudly if tactlessly. "You could hardly be more wrong. Tell me, did you find it difficult to move when you were resting peacefully on the meadow out there?"

"Difficult!" exclaimed Dorothy. "It was impossible. I could no more move than if I had been bound with bands of steel. I do not see how photons could have that effect. They are just light, are they not?"

"Yes they are. But if you *had* been bound with bands of steel, what do you think would actually have been holding you? Why, light again! In any material, steel included, electrical forces bind the atoms to one another, and electrical forces are produced by the *exchange* of photons. These are virtual photons, to be sure, but they are still photons for all that. An electron in one atom happens to absorb photons that have been emitted by an electron in another. Virtual photons can carry both energy and momentum. If one electron absorbs a photon that has been emitted by another, then the energy or

---

[2] Like most technical terms, *real* and *virtual* can be confusing. A real particle has all the energy it needs for as long as it wants. Virtual particles have some sort of energy deficit. They do not have the right energy to provide some or all of their mass and are said to be "off the mass shell." To make things even more confusing, many apparently real particles may be a tiny bit virtual.

momentum carried by the photon has been transferred from the first electron to the second. What a force does is transfer momentum to whatever the force acts upon. Conversely, if you transfer momentum, then you have exerted a force. The photons have produced an interaction between the atoms. All forces, all interactions, are caused by the exchange of some sort of virtual particles. All electrical forces are due to the exchange of virtual photons. Your steel bars are held together by nothing but the stuff of moonbeams."

"Such stuff can only be stuff and nonsense!" exclaimed the girl hotly. "You cannot feel a moonbeam or any light that falls on you, except of course that bright sunlight makes you feel warm. You can certainly feel an iron bar in quite a different way. Light is harmless; an iron bar may not be."

"It is a matter of degree," replied the Acrobat, who was by now juggling the photons so quickly that here seemed to be hundreds of them tossed to and fro. "You might think differently about light being harmless if you were inside a star that had just gone nova. You may safely ignore a few photons, but a fierce blast of them could burn you to a crisp. Photons are given off by electric charges, and in practice that usually means electrons. Real photons, the sort that travel about as the light you see, appear only when electrons are *doing something* that changes their energy—when they transfer from one state in an atom to another or even when they jiggle to and fro in the aerial of a radio transmitter. Virtual photons also are emitted by the charge on electrons, but in that case the electrons do not necessarily have to provide them with energy and do not have to be *doing anything in particular*. There are an awful lot of electrons in the world, and most of them are in some sort of undisturbed state, so they can and will emit virtual photons but not real ones. You might expect there to be an awful lot of virtual photons around, and there are. The flux, the intensity of virtual photons present, is likely to be greater than the intensity of real photons in the blast from an exploding star. On the whole they do not carry the free energy that makes an exploding star so destructive, but what they do they can do intensely, and this includes giving powerful bonding of the atoms within 'solid' iron."

Here the Acrobat seemed to sense that a more concrete demonstration was in order. "If you want to see an illustration of the diversity of particle exchange, try that booth over there." He indicated a gaily striped tent with a poster outside that said:

The Never-Never!

**A game of *infinite* debt
for two or more particles.**

Bet everything you've got
—**and** all that you haven't.

Use your **charge** accounts to borrow as
much energy as *creatively* as you can.

"You might find this instructive," he added as he ushered them inside. In the center of the tent they saw a table, and at either end rested a charged particle with a player seated by each one. A tall scoreboard rose behind each participant.

"Play is just about to begin," explained the Acrobat. "The two contestants control the exchange of virtual particles between the charges, and the scoreboards keep track of the energy they have borrowed for the purpose. Look, one player has just made a rather conventional opening move."

The player had begun with the cautious emission of one long-range photon. This had very low energy and so could exist for long enough to get a good distance away from the charge that was its origin. The energy invested in this had scarcely registered on the scoreboard. On the slick, subtly glowing surface of the table there appeared a wavy line between the two charges, the conventional notation for the exchange of a single photon. The other player responded with a short-range photon of much higher energy. The greater energy made the time allowed to it too short for it to reach the other particle, and instead it contributed to the cloud of virtual particles that surrounds any charge. This cloud became steadily denser as it approached the charge at its origin, where virtual bosons of higher and higher energy could contribute. On the table the photon was shown as a wavy line that turned back to end on the same charge that had emitted it.[3]

The first player then raised the stakes by offering two particles at once, and the second responded with three. Each move was accompanied by the addition of more wavy lines to the diagrams on the table and by a slight increase in the totals on the scoreboards. This steady escalation continued for a little while until a flash of inspiration struck one of the players.

"Hey!" he exclaimed. "You know that we are allowed to exchange only bosons?"

"Yes," was his opponent's cautious reply.

"Well then. If I have a particle–antiparticle pair, that would be much the same as a boson, wouldn't it?"

---

[3] Such diagrams that give a convenient way of representing the exchange of virtual particles are known as Feynman diagrams.

"Yes. So?" still cautiously.

"So I could exchange a particle–antiparticle pair. Say an electron and a positron."

"But you know that only *photons* will couple to electric charge. Your charge has to emit photons and nothing else. That's *all* that it can do."

"Yes, but listen to this. Electrons and positrons have electric charge. Electrons have negative charge. Positrons have positive charge. Both are fine for connecting to photons. So here's what I can do. I'll send off a virtual photon, just like we've been doing so far. Then *that* photon can convert to a virtual electron–positron pair, right out in the middle of nowhere. These will convert *back* to a virtual photon, and that photon will be absorbed by your charge. What do you think of that?"

"Sounds complicated."

"Yes, you're right. It will be too much of a hassle. I don't think I'll bother."

"No, that's not an option. You know the rule: 'Whatever is possible is compulsory.' If you *can* include particle–antiparticle pairs, then you *have* to. You don't have any choice, my friend."

Accordingly, the player did send out a photon that converted to an electron–positron pair and then back to a photon before it was absorbed by the other's charge. This exchange was dutifully recorded as a diagram on the table top, with the particles and antiparticles shown as a closed loop of solid lines among the wavy lines that designated photons.

Thereafter the two players threw caution to the winds and exchanged ever more complicated constructs. They had electron–positron pairs from which the virtual particles in turn emitted their own virtual photons. They could legitimately do this because they carried electric charge. Sometimes these photons were absorbed by an original charge, sometimes by the charge of the other particle in the pair. Sometimes the new photons themselves converted to electron and positron. Dorothy noted that occasionally, but only very close to the primary charge, a photon would convert to a proton–antiproton pair. These were very heavy and drew a *lot* of energy from the account, so they could not get far at all. The situation was getting *really* complex.

Throughout this frantic escalation of the situation, the scoreboards showed an increasing expenditure, or rather borrowing, of energy. This energy was fed into the myriad and ever-increasing varieties of particles that formed a cloud around each charge, as well as a few that made it across to the other particle. A second column on the board, that Dorothy had not at first noticed, showed a steady, and once again increasing, movement in the observable *charge* of the two initial particles, because there were now charged particles in the virtual clouds that surrounded them. The energy that had been invested in the field or cloud of virtual photons and other virtual particles that surrounded each charge was now growing at an incredible and ever-increasing rate. The totals reached the top of the tall scoreboards that stood by each player, and *these* now proceeded to grow and then grow again. They soared up to the top of the tent. Without pause they tore through the canvas. The group of companions rushed outside the tent and watched the scoreboards still rising into the sky. Their tops soared ever up and beyond, vanishing completely from sight.

"Gosh!" exclaimed Dorothy. "When will they stop growing?"

"Never," responded the Acrobat. "There is no limit to the energy involved. The cloud of virtual particles surrounding the charges is infinitely intricate and stores an infinite amount of energy. The amplitudes for the more complex processes are smaller than for single photon exchange, it is true, and the likelihood of such complex processes is small, but there are *so many* options available that the total energy involved just continues to grow. It is far greater than the measured mass energy of the charged particle itself. The great range of possible amplitudes is illustrated in another way by the show over there."

He indicated a small booth with a sign that read *Dance of the 7 Veils*. In front of this sign, a heavily veiled charge was gyrating in a suggestive way. How it did this is quite impossible to describe or even to imagine, but it certainly caught the attention of Dorothy's companions. With great deliberation the charge cast off the outermost veil. This was thin and tenuous, a single low-energy virtual photon that had spread wide around it. More veils followed. Some were thin, single-photon exchanges. Some were more substantial, two or more photons simultaneously. Now and then came an exotic veil bejeweled with a charged particle pair.

It occurred to Dorothy that there must have been more than seven veils by now. She glanced at the sign and saw that it now read *Dance of the 77 Veils*. Still there was no cessation. Veil after veil was cast aside. As the veils came from regions closer to the naked charge,[4] they became more intricate, heavily jeweled with particle–antiparticle pairs and with row after row of fine photon stitching joining them. It was rich, it was exotic, and there seemed to be no end to it. Dorothy tore her eyes away from the dazzling display for a moment and looked back at the sign. It now read *Dance of the 777 Veils*.

"Is there no end to this?" she asked rhetorically.

---

[4] I'm sorry, but that's the technical term for it.

"No, there isn't. There is a never-ending succession of amplitudes for virtual particle emission from the charge. As you get closer to the nucleus, they get more and more intricate and exotic, with greater and greater energies involved. The energy stored in the field and the resultant addition to the mass of the charged particle are infinite."

"How can it be?" exclaimed Dorothy reasonably. "The mass of the particle surely isn't infinite. You told me that electrons were not very heavy at all."

"No, they are not. But that is the mass you see from *outside* the electron, and it is not the same as the mass of the basic particle. You saw previously that the mass of a nucleus was slightly less than the sum of the masses of the protons and neutrons inside it. In the case of a charged electron, you *do not know* the mass the electron would have if it did not have the electric charge that gives the field of virtual particles. You *never* see an electron without its electric charge. The raw mass of a naked electron may be buried far down below the mass that you observe. You have no reason to say otherwise."

"But you said that the energy of the virtual particles was infinite!" protested Dorothy. "The mass of the electron cannot be *infinitely* less than the small value you see!"

"Why not? How do you know what the mass might be? It is rather like speaking of height above ground level. Gravitational potential—the energy that you must provide to raise an object into the air—depends on its height in the Earth's gravitational field. But what is its height? You might say the distance above ground level, above the lowest level to which the object might fall. Now where is that level?"

"Well, that is clear enough," said Dorothy, looking at the level, grassy plain that stretched around, all the way up to the gates of the Emerald (and Ruby and Sapphire) City in the distance. "This is it. It is the level of the ground here. That is as low as you can possibly fall."

"But is it? Just look behind you, but carefully!"

They all looked around and found that, where they had been sure there was level ground stretching in all directions, they were actually standing on the rim of an excessively deep pit in the ground. Just a few feet behind them the earth dropped away with sickening abruptness. Down, down, down; the chasm seemed to fall away without limit. When they crept (carefully) to the edge and looked into the remote depths of this incredible pit, they could see no bottom. There was nothing visible, except perhaps a hint of stars showing through.

"So how far could you fall now? Where is the 'ground level' at which you may say there is no energy potential due to gravity?" He picked up a handful of earth and threw it toward the abysmal chasm. Everyone expected it to van-

ish without trace in the unfathomable depths below, but instead, to their amazement, it spread over the top of the opening and sealed it off from view. There was no telling that the pit had ever been there.

"That is ground level," remarked the Acrobat. "You have no reason now to believe otherwise. As far as you can tell, the ground lies level all about, and there is no reason to suspect, or to worry about, any hidden depths below. If they are there, they are hidden. It is the same with the mass of the electron. What the mass of the naked electron may be you do not know. Neither need you really care. The electron may have an infinite energy debt that is compensating an infinite value for its mass. Even if that should be so, you will never meet an electron that is not discretely clothed in its full complement of electric charge. The 'clothed' mass is the only mass that need concern you," he said firmly.

"The only electron you will ever see is a clothed electron," he repeated. "Any electron—any charge, for that matter—is surrounded by a cloud of virtual photons. Far out the photons are tenuous, with little energy or momentum. Close in to the charge that is their source, the cloud is denser. It stores more energy, and electric charge also, in the form of balanced pairs of virtual particles and antiparticles. When another charge comes into this virtual cloud, it may absorb some of the particles and receive the energy or momentum that they have borrowed. Transfer of momentum from one charge to another is the effect of the electrical force, as I have already said. The cloud of virtual photons *is* the force. The photons are the electric field that surrounds the charge, and when the momentum or energy that they briefly possess is transferred to another charge, that manifests itself as the force exerted by the field. There is no field and no force save that of virtual particles," the Acrobat concluded. "Goodbye," he added abruptly. "I have had too much time already. I must leave while I still have the energy." So saying, he departed with a back somersault and a series of cartwheels that carried him away across the photon meadow.

RENORMALIZATION

The number of possible sets of virtual particles is infinite. Because they are *possible*, they are in a sense actual, and the energy (and charge) associated with them turns out to be infinite. Now, energy is mass, and you know that electrons, for example, do not have infinite mass. This seems to be a paradox.

The way to resolve it is to say that because you know the mass only when the particles are "clothed" in their garments of virtual amplitudes, you may set *that* at the observed value and let the "naked" mass and charge work out to whatever they will.

Surprisingly, when you do this, *everything* that you calculate thereafter works out to be just what is observed in experiments.

Left to themselves, the various members of the group of travelers looked around. They saw that the Building Block Road had re-emerged, leading away from the area where they stood. The path led toward the Emerald (and Ruby and Sapphire) City, which was now clearly visible and no longer looked so very far away. They promptly set off toward it. As they walked, Dorothy mused on all that the Acrobat had said to her.

"I think I can see that borrowing the energy to throw out such a variety of virtual particles might well result in a great energy debt. I think I can even vaguely understand that the energy in the electric field of an electron could modify its mass, because I have been told that energy is mass. What I do not see is why the emission of virtual particles should change the total charge that a particle appears to have," she announced to no one in particular. "Surely I have been told that when a particle is given off, even a virtual particle, the electric charge is conserved." As usual, the Lion treated any question, however implicit or rhetorical, as an invitation to expound. He responded immediately.

"It is indeed because the virtual particle cloud includes charged virtual particles that you may observe a different charge. As you have seen, an electric charge will emit virtual photons, and these may in their turn produce pairs of charged virtual particles. Now it is true that charge is conserved," he continued, settling into a comfortable stride. "The total charge of any pair of particles created must always come to zero. They can have no net charge overall, but their individual charges may be in different positions and can affect the electric field locally. Consider the electrons in a metal. They are free to move around, and so, if there were an electric field within the metal the electrons would move in that field until the fields produced by their own charges had canceled out the field that was originally present. At this point there would no longer be any overall electric force on the electrons, and they would stay where they were. Electrons will always move to cancel any electric field *inside* a metal. That is why you do not get reasonable telephone reception for a mobile telephone inside a metal box."[5]

Dorothy did not own a mobile telephone, but she had heard people grumbling about the effect.

"Does this movement of electrons also affect the field outside the metal?" she asked.

"Near the surface, yes. At a sufficient distance there is not much effect, because the electron charge is balanced by the positive charges on the ions that are fixed in the metallic lattice. The electrons are free to move, the ions are not, but the overall charge of the metal is zero."

"So if you get far enough outside an electron, to a place where there is no charge around, then the electron's charge will not be altered by any charged virtual particles, no matter how the charges may be distributed."

"If you could find somewhere that had no charge, then you might be right. But you cannot do that."

"I do not see why not!" protested the girl. "You just have to find a place where there are no electrons. A vacuum perhaps."

---

[5] Such as an automobile. A car is a metal box with holes in it (such as the windshield), and reception is usually worse inside one. Mobile phones do work inside cars, but that is because the phones are rather sensitive and the holes are quite large.

"I am afraid that would not do. Even a vacuum may have an effect. As it so happens, we are in a good place to illustrate this," he continued as they found themselves walking up a slope. Dorothy saw that the pathway was rising to form a sort of block-built drawbridge into the city. On either side of this causeway there stretched a moat, but it was not filled with water. It was filled with nothing. This was not to say simply that there was no water in it and that the bottom was clearly visible. It was filled with NOTHING: a great emptiness—an absence, even a denial, of matter and content. It was the Vacuum.

Peering down into this pit of nothingness, Dorothy could see within the emptiness a sort of seething whirl. It was a bit like the rolling motion in a liquid coming to the boil. It was a bit like a pond she had once seen squirming with a mass of newly hatched tadpoles. Every part of it was packed with hectic but not quite discerned activity.

"What is going on down there?" she asked, intrigued.

"Well, that is the Vacuum, and within it you see—or almost see—the presence of virtual particles of all sorts."

THE VACUUM

In quantum physics the vacuum is no longer simply *nothing*. It is quite busy.

Virtual particles borrow the energy they need for their brief existence from the quantum energy fluctuations. If there is nothing but the need for mass to inhibit their existence, then they may simply appear, effectively from nowhere. There are some other restrictions that must be obeyed. Electric charge must be conserved, and fermions can be created only along with their antiparticles.

So far we have considered virtual particles as emitted by "real" particles, but they do not really need these. The quantum vacuum is densely populated with such a flickering froth of virtual particles, and this has observable effects.

"But how can that be? You have all said that virtual particles are given off by charges, and I suppose I can accept that. But if that is a vacuum, surely there are no charges there—nothing that can act as a source for such particles."

"Inherently perhaps not. But if there were charges, then they could emit virtual photons, or if there were photons, perhaps they could emit charged particle–antiparticle pairs. Either is possible, with no initial stock of energy, charge, or anything else required. If something is possible, then it is mandatory, as you know. There are amplitudes present for all sorts of virtual particles."

"That's nonsense!" said Dorothy firmly.

"Perhaps so. It is rather like the question of which came first, the chicken or the egg.[6] It remains true, however, that the vacuum is populated with virtual particle amplitudes and that the charges of those that *are* charged may separate to give a *vacuum polarization*. In the same way that electrons and atomic cores may separate in any solid and produce a *dielectric constant* (a factor by which electric fields are reduced within that material), so the virtual charges in the vacuum may separate to reduce the strength of an electric field. Classically, it was thought that there was no reduction of field strength in a vacuum and that no such field could be measured, but this is simply because the field is *always* in the vacuum. Any effect that the vacuum may have is your starting point for the measurements you make, and so any field in the vacuum is not normally detected."[7]

Dorothy still felt that there was something not quite right about all this. There seemed to be something inherently wrong in adding two infinite values that would cancel out to give a fairly small result—and then to expect people to believe in what you had done.

"I don't know," she said. "It all seems so. . . ." She stopped at that point, not quite certain *what* it seemed.

"Arbitrary perhaps?" suggested the Lion. "You may feel that saying that the unseen, naked charge of the electron might have any value, particularly a value that is infinite and negative, is simply sweeping the whole thing under the carpet, as it were. You might well imagine that there is no way of confirming such a radical notion and that it must remain only a pipe dream. There is confirmation, though, at least indirectly. Once you accept that the isolated bare electron may indeed have an infinitely large charge, you can then calculate many other electrical properties. These turn out to be in agreement with the values you observe in experiments—and very accurately too. So you see, when you assign a seemingly quite improbable value to one quantity that is itself hidden from you, you find that you then get the *right* numbers for many things that you *can* measure. It is quite convincing in its way."

There was no arguing with that. Indeed, Dorothy had not so far found arguing with the Lion about *anything* to be very productive. So it was in silence that the group arrived at the Gate of the Emerald (and Ruby and Sapphire) City.

[6] The answer in that case is *the egg*. It was laid by something that had not *quite* evolved into a chicken

[7] There are measurable effects of the virtual amplitudes. One is the Casimir effect, which gives a weak force between two metal plates close together in a vacuum because of the particle amplitudes in the gap.

CHAPTER 9

THE

WIZARD

OF

QUARKS

Dorothy and her companions walked through the open gate of the city and looked around them. Everything was built on an impressive scale and spotlessly clean. The companions were struck at once by the total absence of any visible color, as Dorothy had been when she saw the city from a distance. There were all conceivable shades of grey, but the colors of everything they looked at were completely neutral.

Whichever way they looked, they saw streets that appeared just the same, crowded with a seemingly identical collection of hadrons. They wondered where they should go next. Though the streets were packed with hadrons, they could not see any actual people of whom they might inquire their way. As they were debating their next move, a horse-drawn carriage drew up on the road near them.

"Hop in," said the driver. "I'll take you there."

"But you do not know where we want to go!" protested Dorothy.

"You want to see the Wizard," replied the coachman.

"However did you know?" she answered in surprise.

"Oh, that's easy. *Everyone* who comes here wants to see the Wizard. Come along now. There's no point in wasting time."

They climbed into the carriage, which was fortunately quite large and had room for them all, even the Lion. Soon they were trotting along smartly through the streets. Dorothy was intrigued to notice that, in contrast to the total lack of color elsewhere, the horse was gradually changing through a succession of attractive shades of blue, red, and finally green. Then it began the cycle again. She asked about this peculiar behavior, though with a premonition that the answer was not going to be very helpful. "Ah, Miss," chuckled the driver, "that's the 'horse of a different color' that you will have heard tell of."[1]

Just at that moment they arrived at the entrance to the Wizard's Palace. They got out of the carriage with thanks to the driver, and he drove away, still chuckling. There was again no one about of whom they might ask directions, but because they faced a single high doorway with an impressive corridor beyond, their choice seemed obvious. They passed though the doorway and walked rather nervously down the corridor. That is, three of them walked nervously; the Lion of course strode along in perfect confidence. Together they entered a great audience chamber.

Their first sight of this hall made them pause in awe as soon as they had passed through the door. It had a great expanse of floor and was enormously high. Tall pillars of a pale grey color rose on all sides to a remote arched ceiling of a misty grey that enhanced the impression of remoteness. The room appeared even larger than it was because its curved walls had a cloudy mirror finish that gave distorted reflections of the space within. Beyond the circle of pillars could be glimpsed strange shapes and remote fragments of masonry that served to trick the eye. In this perceived outside world beyond the Wizard's hall, the dimly observed complexity of all that could be seen served to disguise the simple reality present within. In the center of the area hovered a rosette of intense glowing shapes: eight massive hadrons arranged in the hexagonal pattern they had seen in the collector's sample, but grown in stature to an overwhelming presence. The simple symmetry of these hadrons was concealed by many mirrored reflections that confused the mind with too many iterations of the basic order.

"WHO ARE YOU?" thundered a disembodied voice, echoing from all around the chamber.

"I am Dorothy and these are my friends the Observant Scarecrow, the Tin Geek, and the Confident Lion," she replied nervously. "We wish to speak to the Wizard of Quarks."

"I AM THE WIZARD. THE WIZARD OF QUARKS. LORD OF THE HADRONS. MASTER OF THE MOST FUNDAMENTAL PARTICLES THAT MAKE UP YOUR WORLD," the voice thundered all about them.

---

[1] I have included this mostly from nostalgia; it is a joke from the film. The reference to color seemed appropriate.

"That is what I wanted to speak to you about," said Dorothy bravely. "I should like to return to my world, and I have been told that you can help me. Do you think. . . ?"

"**SILENCE!**" bellowed the disembodied voice. The pattern of hadrons that loomed above them changed upon the instant from a rosette of eight to a triangular grouping of ten, such as she had also seen earlier in the collector's drawers. This greater number of hadrons rose higher and appeared even more ominous than before as it floated oppressively above their heads. The echoes of the tremendous voice rolled around the chamber, and the very walls seemed to shake. What actually *did* shake was a curtain that screened a corner of the room, and Dorothy fancied that she saw movement behind it. She went over and lifted the curtain aside.

"PAY NO ATTENTION TO THE LITTLE MAN BEHIND THE CUR-TAIN!" roared the great voice that filled the room. That was indeed what Dorothy had found when she moved the curtain: a little bald-headed man pressing buttons and speaking into a microphone suspended before him.

"Who are you?" she asked.

"I AM THE GREAT AND MIGHTY . . . Wizard of Quarks," he finished in a small voice, turning away from his microphone to face her.

"Are you really the Wizard?" she asked in disbelief.

"Yes, I am afraid so. I am all there is, apart from the quarks, of course."

He gestured vaguely at a number of small, colored objects that clustered around his feet in the cramped space. At the same time, the great, variegated array of hadrons, which had so oppressed them from above their heads, came tumbling down and left nothing but a collection of the same little, colored quarks scurrying around the floor.

Dorothy looked at her discovery more closely. Now that they could see him clearly, he was revealed as a small, elderly man with a bald head and side-whiskers. He was distinctly stout and currently looked rather embarrassed. If he was not the hadron collector they had met earlier, then he was his identical twin, though now he was dressed more formally in a frock coat in place of the track suit they had seen before.

"Haven't we met?" she asked. "Didn't I see you collecting hadrons in the Kingdom of Cern."

"Well, yes," replied the Wizard, looking if anything even more embarrassed. "It is a hobby of mine, collecting hadrons. I am always on the lookout for one even rarer and more exotic than those already in my collection. This may seem a bit pointless to you, inasmuch as all of the hadrons are only groups of quarks after all. That is why I dress things up a bit here. To make it all appear more mysterious and exotic you know."

"But are they really all just groups of quarks?" asked Dorothy. She had no particular reason to doubt the information, but some reply seemed in order.

"Oh, yes. The baryons—those are the fermions, you know—are made up from groups of three quarks. Mesons are strongly interacting bosons. They are combinations of a quark and an antiquark."

THE QUARK COMPOSITION OF HADRONS

Hadrons are those particles that interact with one another through the so-called *strong interaction*. There are many different types (as noted in the text), both fermions (called **baryons**) and bosons (called **mesons**). They are all constructed from a small number of more basic particles called quarks. Baryons are combinations of three quarks; mesons are combinations of a quark and an antiquark.

We do not at present know of any particles more basic than quarks.

"Are there also antiquarks then?"

"Indeed there are. Every particle has its antiparticle, though in some cases they are, perhaps surprisingly, the *same* particle. The antiparticle of a photon is still a photon, for example. All fermions have distinct antiparticles, particles that are *antifermions*. Quarks are fermions, so there are antiquarks that are quite distinct.

Let us get out of this silly hall now. Its only purpose is to impress, and that does not seem to have worked very well."

They made their way out of an arched opening in one corner of the vast audience room. This led into a paneled corridor with a mosaic-paved floor.

"It seems strange that all those exotic arrays of hadrons should be made up from just a few quarks, if that is the case," remarked Dorothy.

"Remarkable it may be," replied the Wizard, "but it is nothing new, you know. You find that the world is composed of such layers, with great complexity in one layer being built up from a few items in the layer below. It is a case of wheels within wheels, you know. That's it in a nutshell."

"What is in a nutshell?" asked Dorothy in some confusion.

"Why, another nutshell. This is a sequence that repeats and repeats and repeats. The world looks very complicated and diverse, and then you find a new level. This new level of matter seems to contain only a few objects, and you discover that these more basic objects combine with one another to produce all the complexity that you saw up to then. Suddenly the world seems exquisitely simple. Then you find that there are more of these basic objects than you thought at first. Then you find still more, and it is too much. The simple picture has become complicated again."

"Oh dear," said Dorothy. "What happens then?"

"Well, eventually you find that there is a new level. One with just a few objects from which everything you knew is made up. A new level that *seems* basic and wonderfully simple. The general scheme is illustrated over there," he finished abruptly.

Dorothy saw that the corridor they were following had opened out to give a circular rotunda, with a shallow, domed ceiling surrounded by a ring of spotlights that pointed toward the center of the floor. The floor here was decorated with a circular pattern, and from its center rose an elaborate frame that held a great spherical sculpture. The sphere filled most of the circular area and rotated slowly in its mounting. It was elaborately carved, and among the writhing shapes that covered its surface could be seen the forms of all sorts of animals. There were cows, dolphins, people, dogs, cats, and mice. Between the larger forms there was finer carving: tiny detailed insects and miniature creatures extending down to level of bacteria. Some of the carving was incredibly fine, indeed microscopic. It extended, without any break in the depth of available detail, right down to a depiction of the double-helix structure of DNA. Dorothy was not sure how she could be aware of this, given that such intricacy was well below the limit of her normal eyesight. Nonetheless, she knew that the amount of possible detail extended uniformly, without a break, right down to this level.

The surface of this great sphere was pierced here and there with gaps in the otherwise continuous carved surface. Through these could be seen an inner spherical shell, this one decorated with representations of the atoms of which all the world above was constructed. Dorothy saw hydrogen, nitrogen, oxygen, and of course carbon. There were other atoms too; iron, silicon, phosphorus, gold. The inner sphere rotated relative to the outer, and as it turned, more atoms could be seen through the gaps in the outerward appearance of things. Tungsten, titanium, platinum, uranium; more and more atoms of different types appeared until there were around a hundred of them, together with their various isotopes. Even this shell, though much simpler and less diverse than the outer layers, was beginning to appear rather complicated.

This inner shell was itself pierced with gaps through which deeper structure could be glimpsed, and as the construct turned, the watchers could see yet another inner sphere. This first appeared to be very simple and basic. Scattered all over its surface were symbols for electrons, and these, accompanied by protons and neutrons, provided the basis for all the atoms shown in the shell above. As this inner shell turned within the outer, however, it soon became clear that the protons and neutrons were not the only hadrons present. Dorothy and her friends also saw hyperons, Δ and N* resonances, π and K mesons, ρ, Ψ, Ω and many others. The entire Roman and Greek alphabets had been called into service and were still inadequate. They had to be enhanced by subsidiary labels such as π^+, π^0, π^-, in order to uniquely identify each member of the hadronic assembly. The number on show was much greater even than the number of atoms in the shell above, and so this level also began to look rather unsatisfactory as a final simplification of the world.

Indeed, this shell too was pierced with irregular breaks through which yet another level could be perceived. This last hidden sphere carried the colored representations of the quarks. Perhaps surprisingly, these were shown still in the company of the electron and the few other leptons. The electron, the first particle ever to be discovered,[2] took its place along with the quarks in the assembly of what were apparently the ultimate basic building blocks of nature.

This final inner shell had no breaks in its surface, and no further inner structure could be seen.[3]

[2] The electron was observed and identified as a particle in Victorian times, by J. J. Thomson in 1897 in Cambridge (England).

[3] As far as is known at present, the quarks and the leptons (together with the intermediate bosons that carry the interactions between them) are the final basic components of Nature. I have to admit, however, that the same was thought earlier of the proton and neutron.

"This is one of my favorite works of art," said the Wizard. "It is called *Nature*. It illustrates the successive layers of physical reality, the items on each layer being compound structures built up from the elements on the layer below. With my quarks I have managed to get in on the ground floor. The quark stops here, you might say. Come along and I shall show you more of the nature of quarks."

They circled around the intricate representation of *Nature* and followed the Wizard down a passage that left the rotunda opposite to the passage by which they had entered. This was considerably less flamboyant than the passage they had been in before. The walls were painted in a dull, dingy yellow, and the floor was of bare planking with a thin strip of worn carpet down the center. A short distance along the corridor stood a vending machine in an alcove. It bore a label that read *"Quarksicles: Three colors! Six flavors!"* and had on its front a row of buttons that were labeled "color" and "flavor," respectively. The Wizard asked Dorothy if she would like to make a selection. "You must select three flavors," he instructed her.

"Do I have to?" she asked.

"Yes" he said, so she pressed a button marked 'u', another marked 'd' and then, for no very good reason, the one marked 'u' again.

"Now do I select the color I want?" she asked.

"No, there is no need. You will get three quarks anyway, one of each color. You have no choice in the matter."

As he was speaking, the machine disgorged a small group of three colored objects: one red, one green, and one blue. Strings of some sort tied them to each other.

"You made a good, if somewhat conventional, choice. That is a proton you have there. It is made from a group of three quarks, as are all the strongly interacting fermions, and in the proton's case it is made from two quarks of 'up' flavor and one of 'down' flavor."

"Up and down aren't flavors!" protested Dorothy.

"Oh yes they are! At any rate, they are for quarks," was the Wizard's reply.[4] "Both the proton and the neutron are made from quarks of the up

---

[4] Quarks are said to have color and flavor. These are just names and have nothing to do with the color or flavor of anything you might see or eat. The "colors" are called red, green, and blue, but these are *still* just convenient names that have nothing to do with ordinary color. The colors that we see depend on the frequency of visible photons that enter our eyes. The names of the quark "flavors" do at least emphasize that we are not speaking of ordinary flavor. They are called up, down, strange, charm, top, and bottom—no strawberry.

and down flavors. The neutron is pretty much like the proton in that it also is made only from 'up' and 'down' quarks, but it contains only one of the 'up' flavors and two of the 'down' flavors. These two types of quarks are sufficient to give, on their own, the main constituents of all the nuclei of all your atoms. There are other types of quark, but they appear only in more exotic hadrons such as you do not often see. Life, as they say, is full of *ups* and *downs*."

"Why did I have to have three quarks anyway?" asked Dorothy.

"So that you can have three different colors, of course. That way you can get a colorless particle, and it is such colorless particles that make up your world. The three-quark combination gives you the baryons. The other way to make a colorless particle is to have a colored quark combined with an antiquark of the appropriate anticolor. This is what you have with mesons. That combination is color-neutral, in the same way as a positive and a negative electric charge give an electrically neutral object. You know that the atoms of which your world is made are electrically neutral."

"Yes, I have heard that," said Dorothy. "Is this color you speak of something like electric charge, then?"

"Well yes it is, more or less," replied the Wizard. "Color is. . . . "

"Stand aside, mister. Color is *my* department." The interruption came from a new figure that they had not observed before. No one had seen her come. It was as though she had been there unnoticed all the time—as of course she had, because the fundamental interactions are always present. She was fairly petite and was dressed in black like the other Witches, but on her it looked good.

Her cloak was a short black cape, and her black dress, which fell only to mid-thigh, was complemented by high black boots. She wore a conical hat like the other Witches, but this was a

R Gilmore

tiny affair almost lost in the exuberant curls of an Afro hairstyle. Waves of green, blue, and red washed continually across her features.

"I am the Witch of Color," she announced to nobody's great surprise. "I look after the color forces that bind the quarks to form the hadrons. You asked whether color was like electric charge. It is, but in a more liberated form. I have greater freedom, for there are three types of color charge. These are usually called red, green, and blue, though they have nothing whatever to do with the mundane colors you know. They are the secret colors that lie behind the colorless world you see."

"My world is not colorless," stated Dorothy hotly. "There is color all around me in Kansas, though admittedly with a bias toward green."

"Pay attention, sister," directed the Witch shortly. "I'm not talking of the colors of light, but of quarks. Your fat friend here has just told you how the color charges cancel in much the same way as EM's electric charges, but I make a better job of it than she does. EM is a bit limited because she has only one type of electric charge, whereas I have three different color charges: red, green, and blue."

"Aren't there two types of electric charge—positive and negative?" the girl protested.

"No, they are not two different types. One is charge and the other is anticharge."

"Which is which?" Dorothy inquired.

"Whichever you like. Charge and anticharge are opposite. Positive and negative charges are opposite. You can say whichever one you like is the charge, and the other would consequently be called the anticharge. The point is," she continued firmly, "that three types of charges offer more ways of making a neutral combination than you have with just one type of electricity. One possibility is still to have charge and anticharge of the same type, just like the electrical case. Mesons are made like that, they are made from a quark and an antiquark, and they may have the colors red and antired, or green and antigreen, or whatever.

"Do the antiquarks have anticolor then?" guessed Dorothy.

"Yes," said the Witch shortly. "There is another way of making a neutral whole, though," she continued, " and that is to have three quarks, one of each color. Red, blue, and green combine to give no color at all. That is the way this sort of color works. The most dramatic effect of having three sorts of charges comes when you consider my exchange bosons."

"What are they?"

"They are the carriers of the color interaction. EM has her photons, and they transmit the electrical interaction. I have gluons, and they do the same

for color. There is one big difference. Intermediate bosons that carry an interaction between two charges carry a mixture of charge and anticharge. In
the case of photons they carry a mixture of positive and negative charge, and
that means they have no charge at all. Photons are electrically neutral, so the
photons themselves do not interact electrically with one another. It is a different case with gluons."

"Oh?" said the girl. The Witch had paused so she felt some comment was
expected, but she couldn't really think of one.

"Yes. The gluons also carry a combination of color charge and anticharge, but now they do not have to be opposite. Their color does not have
to cancel. Because there is more than one type of color charge, you might
have red and antigreen present in one gluon, or blue and antired in another.
Gluons themselves carry color, and this has *very* long-range consequences.
Come and see."

They walked along the worn carpet down the center of the corridor and
turned left through a plain, no-nonsense door. They found themselves in a
long, narrow room with a number of more or less traditional toys scattered
around it.

- There was a spinning top, but this top went on and on spinning at exactly
  the same rate, without ever slowing down. Furthermore, from whatever
  direction you looked at it, its axis seemed to lie along *that* particular
  direction.
- There was a board game with the pieces set out. It looked something like
  "snakes and ladders" but involved electrons climbing up to or sliding back
  down from various atomic levels.
- There was a pile of building blocks, and on closer inspection, each block
  was revealed to be built up from blocks smaller still, and these smaller
  blocks were in their turn divided. . . .
- There was a large box, completely featureless apart from a single door in
  one side.

The Wizard opened this door and stood politely aside for the Witch and
Dorothy's group to pass through before he followed them in. Inside there
was nothing. That is not to say that they found themselves inside an empty
box. There was *nothing*. They were in a totally featureless region, void and
without form.

"This is a toy universe," remarked the Wizard brightly. "It is very useful.
Every theoretician needs one. It gives you somewhere to look at things without the complications that arise in the real world. If you want to examine two
quarks on their own, you need a toy universe to put them in. In the real

world the quarks are combined into hadrons. The hadrons are mostly in atomic nuclei, which are themselves in the center of atoms. The atoms in turn are combined to provide all the molecules and compounds from which your world is made, and you *know* just how complicated the world can get at times. To keep things simple, you need a toy universe of your own—one like this."

"OK," said the Witch crisply, moving in front of her audience. "To show the differences between electrical and color forces, I think we had best look at the interaction between two charges. First we shall see what happens with electric charges. I think you ought to do this bit," she added, turning aside to address the tall figure of EM. No one had noticed her arrive. It was as though she had been there all the time—as of course she had.

EM moved forward to stand in front of them. "Observe an electric charge, such as would be carried by an electron," she began. A single electric charge appeared in front of her, floating in the otherwise empty universe.

"Around this charge is a flux of virtual photons," she continued, bringing her remarkable staff down with a ringing crack. (This was all the more remarkable in that the universe contained nothing for the staff to strike.) From the charge there sprang a tracery of fine lines, radiating outward like the spokes of a wheel and extending out to infinity. As they got farther from their source, they became more spread out and diffuse, thereby demonstrating how the photon amplitude became less intense at greater distance and how the strength of the electrical interaction fell off with distance.

"Now observe how the presence of a second charge, of opposite polarity such as on a proton, will affect the field," commanded the Witch. Some way off a second charge appeared in the toy universe. This charge had flux lines spreading out from it, just like the first. The lines from the two charges overlapped, and where they did, they combined, as one would expect for amplitudes. The overlapping flux lines from the two charges blurred together to give a single set of lines that departed one charge, curved widely between the two, and then converged onto the other charge. Away from the line joining the two charges, the flux lines swept out in great curves to spread far and ever wider away from the charges that were their source and also their destination.

THE ELECTRIC, OR PHOTON, FIELD BETWEEN TWO OPPOSITE CHARGES

In the electric field between a positive and a negative charge, the flux lines followed by the amplitude of virtual photons spread out and become more diffuse as you get farther away from the charges. As this probability amplitude for the virtual photons becomes more diffuse, the field becomes weaker as the charges move away from one another.

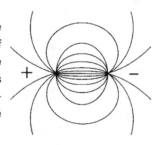

"You can see that in the case of my photons, there is no electrical interaction between them. Therefore, the lines spread apart, becoming more and more diffuse the farther away you get," remarked EM. "The flux that begins on one charge ends on the other, so the electrical effects are largely contained within a region close to one or the other charge. At large distances the effects of the two charges cancel, and so you find that over a sufficient region, the pitiful field of the Witch of Mass becomes dominant." EM sniffed condescendingly before going on.

"Outside a neutral atom the electric fields of electrons have little effect at any distance. For atoms that are close to one another, the electric fields do not *quite* balance, because the electrons and the positive nucleus are not in exactly the same place within an atom. This slight displacement of the opposite charges, and the consequent inexactitude in their cancelation, create the force that binds atoms together in a solid. Not only is the strength of an iron chain provided solely by the virtual photons of the electric field, but it comes just from the small effect left when opposite fields fail to cancel. Because of the cancelation between opposite charges, there would be little sign of electricity in the wider world beyond the atom were it not for the fact that my bonds, though strong, are not absolute. They *can* be broken."

The Witch of Charge pointed her finger at the more distant charge, and a high-energy photon flashed from her fingertip, striking the charge and giving energy to it. This charge promptly rushed away from the other, although the electric field sought to restrain it. As it moved away it lost energy, fighting against the electric force. It slowed, but as it moved farther away, the force became steadily weaker and so the rate at which the charge lost its energy went down. It was a case of diminishing returns, or rather losses, and it was the weakening field that lost the day. The charge moved off to vanish into the distance of the toy universe. This had seemed quite small before they entered it from the nursery, but inside it stretched to infinity, as even a toy universe must.

"The two charges are now free of each other, as happens when an electron is torn from an ionized atom. Between such charges the full electrical interaction may be felt over long distances, and it is the few such liberated charges that keep the banner of electromagnetism flying high amid the populist uniformity of gravity. These free charges produce electrical effects that you can detect even in your large-scale world. Free-roaming electrons that have torn loose from atoms (atoms that remain anchored firmly within metals) give the electric currents that flow within your electric lights and through your TV tubes."

"Right!" the Witch of Color broke in. "Now, *my* interaction is stronger yet and has a reach at least as long as EM's, but you probably have not seen any sign of it at all. It is *because* my interaction is so strong and reaches so far that you do not see much sign of it."

"That seems to be a complete contradiction," protested Dorothy.

"Not if you think about it, sister," returned the Witch. "You are aware of gravity because all the mass in the world works together to reveal it. There is no cancelation there, no internal conflict to limit the effect at long distance.

You have recently learned that despite the much greater strength of the electrical interaction, you are aware only of the effect of those few charges that have broken loose from neutral atoms. Most of those are still contained within metals where their effect is canceled at large distance by the presence of fixed positive ions. Charges that are truly free are few indeed, but it is these that give all the electrostatic effects you ever see.

"In my case there *are* no such free color charges. The quarks that carry the color cannot break free—ever! Let me show you. Observe what happens with a colored quark." Once again a charge materialized, floating alone in the apparently empty universe. This one was of a vivid but fluctuating color. From it led a fan of two-colored lines.

"There you see the quark and the flux lines of the gluons that it emits. As with the photons, the gluons carry charge and anticharge. However, as I have told you, the two do not cancel because there is more than one type of color. The gluons themselves are colored, and they can emit other gluons."

Along the gaily colored lines they could see little dots, like tiny colored spiders, from which spun out new colored lines. These formed a mesh that connected the gluon flux lines themselves.

THE COLOR, OR GLUON, FIELD BETWEEN TWO QUARKS

The flux lines of the gluon field between a quark and an antiquark are confined to a narrow bundle. The color charges carried by the gluons give extra interactions within the gluon amplitudes. They do not spread out widely, as does the flux of virtual photons in an electric field. The energy remains fairly constant along the bundle, and so the force between the two particles remains fairly constant as they move farther apart.

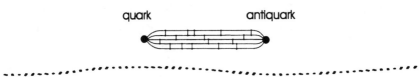

A myriad of these extra lines could be seen, the cross-strands in a spider's web of flux lines. Under the influence of this cross-stitching of color lines, the flux lines from the original colored quark were pulled together and bound into a tight bundle. It was this tight bundle that joined the quark to an antiquark that could now be seen some distance off. There were no flux lines spreading wide on either side, as they had for the electric charges. The entire gluon amplitude, the amplitude that formed the color field between the quark and its companion, was confined to this narrow, multicolored rope.

"You see how the color interactions between the gluons themselves have stopped the flux lines from spreading out and becoming more and more diffuse as they get farther from their source. Because the gluons act as new color sources, the interaction is always strong, and the density of energy in the field is much the same at any distance. See what happens in this case when you give energy to the antiquark."

She raised her hand and hurled something at the antiquark. The impact drove the antiquark away from its tethered companion. As it moved off it lost energy, because it moved against the force that bound it to its fellow, but now the force did not weaken with distance. The antiquark lost energy at a steady rate and moved away only until it had lost all of the impulse it had been given. Then it stopped, still bound as tightly as before.

"No matter how much energy you give to an antiquark or quark that is bound, it will not break free, because the gluon field can absorb *any* amount of energy. As I said, the interaction is so strong and of such long range that you cannot get free color charges."

"Can these bonds never be broken, then?" asked the girl. "Surely something must give eventually."

"Yes, you are right there. Something does indeed give, and the bonds can be broken in a way, but not as you might expect. It is not a way that releases a free quark with its color charge exposed. When you give more energy to a bound quark or antiquark, it will move away and energy will go into the gluon bond. You can never provide enough energy to pull the two apart, but you can impart enough energy to create the masses of a new quark–antiquark pair."

She raised her hand again and hurled something even more forcefully than before at the relatively distant antiquark. Again the impact drove this antiquark away from its tethered companion, and the multicolored rope between the two stretched even farther than before. This time, however, the antiquark was not brought to an eventual halt. Instead the string parted, giving up some of its stored energy to create a new quark–antiquark pair. The antiquark provided a new termination to the string, but this antiquark was *not* carrying the momentum or energy of the other, and the string snapped back to its original length. The quark member of the newly created pair had bound closely to the original antiquark with a short thread of colored gluons between them. It had formed a meson, a pion, and this pion rushed off, carrying the excess of energy and momentum.

"You see now," remarked the Witch of Color, "that instead of separating an isolated quark, with isolated color, all that I have accomplished is to create a pion in addition to the original particle. This is a common enough event. I am sure that you saw many pions being created when you visited Cern. If there is a great deal of energy available, many pions are likely to be produced, and other particles as well, but what you will *not* see is the release of free quarks."

"It is not fair to say that my photons can give no interaction when they are between the primary charges," protested the Witch of Charge. "As you are well aware, they may give rise to charged particle–antiparticle pairs, and these in turn may create other photons."

"Yes, but such behavior can happen only very close to the original charges," returned the Witch of Color. "The charged particles all have mass, and a great deal of energy is required to produce this mass. Virtual photons that have purloined enough energy to manage this are restricted to a region *very* close to the source charge. At reasonable distances from the source charge, the virtual photons cannot obtain enough energy to create even the lightest charged particles. They cannot exchange them over significant distances. Consequently, the large-scale behavior of the flux lines—the *lines of force*—is just as you previously described it, with no linking of the flux lines to one another.

"I admit that at very short distances, as seen by particles of very high energy and momentum, the electric charge becomes less effectively *screened* by the charges of such virtual particles," continued the Witch of Color with a nod toward EM. "Your electromagnetism becomes *stronger* at such very long distances. The opposite is true for color. The enormous color activity that occurs at large distances from the primary charges, between the gluons themselves, dominates completely, and this means that the force actually becomes stronger at large distances. At short distances it is relatively weak and not unlike electromagnetism, but over long distances it cannot be challenged.

QUARKONIUM

The color force is peculiar in being strong at long distances (low energy), but relatively weak at short distances. This is called *asymptotic freedom*. When such short distances are probed by high-energy particles, it is found that closely coupled quark–antiquark combinations behave rather like simple atoms, with a similar set of excited states. Such a system is sometimes called *quarkonium* because it seems remarkably similar to *positronium*, the electrically bound state of an electron and its antiparticle, the positron.

"And that," she finished, turning to Dorothy, "is how the color forces are like, yet not like, those of electricity."

"Yes, but now you should be on your way," broke in the Wizard abruptly. "You asked me earlier how you could get back to your own world. You have farther yet to go. Come along," he commanded, opening a door that appeared as a brightly lit rectangle in the featureless void. He led Dorothy and her group out through it. The Witches did not follow because they were of course already outside, having been there all the time. Dorothy and her companions filed out and found themselves once again in the nursery.

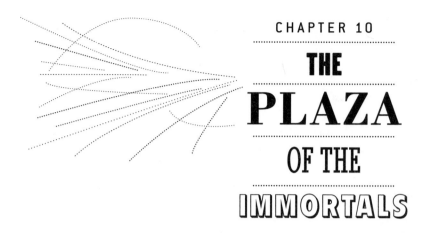

CHAPTER 10

THE
PLAZA
OF THE
IMMORTALS

They looked around the nursery. They saw the various toys that had been there before, and over at the far end, they now observed a number of babies in cots.

"Whose are those children over there?" Dorothy asked the Wizard. "Are they yours?"

"Oh dear me, no! This is the *New Notions* nursery, and each one of those is the brainchild of some theoretician. They wait here as little newborn hypotheses, some of them scarcely more than a passing thought. They have to remain here, safely out of circulation, until someone comes along with the experimental evidence that is necessary to validate them. When actual measurements have adequately confirmed them, they can leave as full-fledged theories. After that, a theory has to mix into the rough and tumble of experiment, and it can be *very* rough at times. It can happen that theories of quite long standing come up against observations that they cannot explain, and then I'm afraid they just have to go. It is sad, but that is how things are."

"Where do they go to?" asked Dorothy curiously.

"They are pensioned off to *Dunexplainin,* a retirement home for Distressed Theories that have fallen on hard times. You will pass it on the way out of here. Those theories that successfully survive every bruising encounter with the harsh light of experimental reality, however," continued the Wizard with increasing enthusiasm, "go on to become *Laws of Nature.* They are the Immortals, the solid foundations of our understanding of the world. You will probably see memorials to some of them in the *Plaza of the Immortals* as you go on your way."

On their way to the exit they passed the row of cots. In each there was a little hypothesis, dressed in a diaper and waving chubby limbs in the air as it explored the new physical world it was called upon to explain. The end of each cot bore the name of the hypothesis within. One cot was different.

R Gilmore

This cot was filled to capacity with the bulky form of a massive teenager, still wearing a diaper but glowering around from under a thick thatch of hair. His shoulders caused the sides of the cot to bulge outward, and his legs dangled over the bottom of the cot, which bore the improbable name of SUSY.

"Who or what is that?"

"That is Supersymmetry, more familiarly known to his followers as SUSY. Supersymmetry is a notion much beloved by many theoreticians. The idea is that at a basic level and at a high enough energy, there is a deep symmetry such that even fermions and bosons look the same. Symmetries such as this are very valuable in physics. Their existence may discourage the incipient anarchy that could come from an infinity of combining amplitudes and keep the amplitudes under tight control."

"So why is he still here if it is such a good theory?"

"Well, this concept required that every particle—fermion or boson— have a companion of the opposite type. An electron, for example, is a fermion. Supersymmetry would require that it have a boson companion called a 'selectron.' A photon, on the other hand, is a boson. It is required to have a fermion companion called a 'photino.' The problem is that each and every particle is required to have a companion of the opposite type, and so far not a single one of them has been found."

"I am surprised that he is still around then," the girl remarked.

"Well, his parents and sponsors are remarkably fond of him. They simply will not give him up. So here he is, waiting for the discovery of even one companion particle. It encourages patience."

With these philosophical words they arrived at the door of the nursery. This door led directly into a paved street, and there the Wizard turned to face them.

"I must leave you now, or rather you must leave me. My place is here, and I cannot accompany you on your further journeys. Before you go, however, I can make each of you a gift to help you in the future. For you, my tireless tin companion, I have this gift" he announced to the Tin Geek. Picking up a canvas sack that was conveniently positioned by the door, he reached within it and took out some metal component. "This is a surge filter for your power supply. It may help to avoid any further unfortunate crashes of your system.

"And this is for you, diligent observer," he continued, turning toward the Scarecrow. Reaching again into the sack, he withdrew a handsome bound notebook with a gold-tipped pencil attached. "You need worry no further about any absence of brains in your straw-filled cranium. This will allow you to make notes of all your observations. Any student will tell you that it is much better to have a full set of notes than any collection of facts in your brain.

"And for you, my prolific friend," he added, turning now toward the Lion. "Your informative output has been so extensive that you can never readily find the exact section that you require. I have for you a packet of *déjà lit* bookmarks. These are akin to the *déjà vu* feeling that one sometimes gets. The difference is that *déjà vu* is a memory of something that you have not actually seen, but is now happening. These bookmarks, however, allow you to locate a passage that you have not actually read but now require. You should find them very useful."

"And what about me?" asked Dorothy. "I do not think there is anything in that bag that will allow me to get back home."

"No, there is nothing in the bag," admitted the Wizard, "but I do have something for you. It is something that can be very valuable, though it is not always well received. It is advice.

"You must go on," he said, firmly and a little sadly. You must continue your journey of discovery as far as you can go. You must continue on your way, traveling beyond all hope and perhaps past all reason. You may have to continue on and on, even through the Great Experimental Desert, but you will reach your destination in the end," he proclaimed confidently. "Have a nice trip. Goodbye," he finished. He shook them all warmly by the hand (or paw, as appropriate) and closed the door firmly behind them.

After he had gone they realized they had not been given any definite directions, but the Lion confidently stated that they should turn to the left, so they did. A little way down the street, the Scarecrow observed a sign that read *Dunexplainin Retirement Home*." This was in front of a tall, severe building whose appearance somehow subconsciously whispered, "Victorian Institution."

"That's the place the Wizard mentioned," exclaimed Dorothy. "Let's go in there."

Obediently, because nobody had a better idea, her companions all followed her through the tall paneled door, across a dark paneled hallway and into a large room. This was not much more cheerful than the hallway. It gave an overall impression of shadows and silence, for heavy curtains were drawn over the tall, leaded windows, and thick dusty carpets absorbed the sound of their footsteps across the floor. Scattered around the room were massive leather armchairs, with winged backs that served to hide their occupants until one was really close.

In the first chair they came to was an ancient, wizened figure. He seemed scarcely aware of their presence, or indeed of anything else, and sat with is head sunk down on his chest, fondly murmuring "Ha! Phlogiston!"[1]

"I think he must have been here for some time," whispered the Scarecrow.

"Not for as long as that group over there, I fancy," the Lion asserted.

The others looked across where he indicated and saw a group of men with impressive beards. These gentlemen were wrapped in what appeared to be white sheets borrowed from the institution's beds and were manhandling a massive slab on the top of which was a map of some sort. It resembled the area around the Mediterranean Sea. Dorothy recognized the distinctive toe of Italy from her geography lessons. The men were trying to balance this slab on the backs of four little elephants but were having difficulty. Every time they had it almost in position, one or another elephant would slip squealing from the curved shell of the enormous turtle on which they were all standing.[2] One of the elephants looked distinctly pink—perhaps from its exertion, Dorothy thought.

This straining group looked much too busy to interrupt, so the companions continued their progress across the room. Their attention was caught by the occupant of one of the chairs, though perhaps *occupant* is not quite the right word because he was not sitting in the chair. He was hovering some distance above the seat, suspended in the air by pulling mightily on the laces of his own boots.

---

[1] The phlogiston theory was an early explanation of combustion. When something burned, it was thought to give off phlogiston; you could see it coming out as flames, after all. The theory suffered from the drawback of most theories that looked very plausible but were eventually abandoned. It wasn't true.

[2] In classical times it was obvious that the Earth was a pretty solid affair and consequently something must be holding it up. One suggestion was that the Earth was carried on the shoulders of a giant called Atlas. Another notion was that the Earth was a large flat plate supported by elephants that stood on the back of a giant turtle. You were not supposed to ask what the turtle stood on.

R Gilmore

"Who are you, might I ask, and how do you do that?" Dorothy said without any preliminary introduction. It was rather forward of her, but her curiosity was aroused.

"Why, I am the Bootstrap Theory. I was a hypothesis about the nature of the basic elementary particles. My contention was essentially that there is no such thing."

"I do not see how that could be," protested Dorothy. "When I visited the Kingdom of Cern, I myself saw them being created. Large numbers of them."

"Well, that was the problem, really. When there seemed to be only a few 'elementary particles,' then you could believe that they were elementary. When the number of different particles got into the hundreds, however, it no longer seemed very plausible that each and every one was truly *elementary*. It seemed likely that they were made out of *something*, but what?"

No one even began to answer *that* question, so the Bootstrap Theory

continued. "When you consider the
strong interaction that acts on every
hadron, you realize that every one is
closely surrounded by—indeed, contains
at its very heart—a combination of amplitudes for all sorts of virtual hadrons. It
is in effect *made up* of these. There are all
sorts of hadrons present, *including* the initial hadron you were considering. In this
sense a hadron is composed of all the
hadrons, *including itself!* You may have
heard it said that someone could scarcely
contain himself. Well, a hadron is adept
at containing itself, together with many
other hadrons. I maintained that everything is just a manifestation of the same
basic 'stuff' or 'field' or whatever. There

R Gilmore

must be some *self-consistency* requirement that results in the group of particles
you actually observe as the only possible selection of particles that could be
composed from those same particles themselves. You could then predict the
properties of all the particles that exist because they are the only ones that
could exist, you see." His speech, which had started off calm and almost subdued, had by this stage risen to a missionary intensity.

"But aren't the particles you speak of all made out of quarks?" asked
Dorothy, who was feeling a little confused by all this.

"Well, yes," he answered, his voice becoming flat once more. "Why do
you think they put me in here? All the experimental evidence that supports
the actual existence of quarks cut the ground from under my feet." He
looked down to where his feet dangled some distance above the ground. "If
it had not been for that," he cried, "I should be celebrated in the Plaza of the
Immortals by now."

"Oh yes," interrupted Dorothy quickly. She was becoming rather good at
seizing on suitable openings. "We need to pass through the Plaza. Would you
tell us how to get to it please?"

"Just go out the front door and proceed down the street. You can't possibly miss it," he answered despondently, releasing his grip on his laces and
sinking down onto the chair in exhaustion.

They left him, sadly huddled in his enormous chair, and made their way
back out to the street. They had not gone far when the street opened out
into a broad paved square, studded all over with monuments of some sort.

This was obviously the Plaza. The band made their way across the pavement, stopping briefly to look at one or two of the monuments as they passed.

They saw a statue of a man in seventeenth-century dress. Around his head orbited a number of small objects. Dorothy couldn't be quite sure, but they did look rather like apples. The plinth bore the inscription "Theory of Universal Gravitation." Another statue was that of a man in more modern attire, with a drooping moustache, a great shock of untidy hair, and sad-looking eyes. Around this monument the space was twisted and contorted into mounds and deep hollows, and this plinth was inscribed "Theory of General Relativity."

Another monument looked strangely indefinite. There was a figure on this also, though it was far from clear that it was one figure. It seemed to be many, all blended together. There was a Danish soccer player, a minor French aristocrat, a wise-cracking American, a long taciturn Englishman, a wavy Austrian, and a tabular German.[3] There may have been more—the probabilities for discerning others in the superposition were high. Permeating the whole region was a lingering sound of intense argument. The plinth said simply "Quantum Theory."

Still farther on they came to a plinth that carried no statue and bore the inscription "Gauge Theory." Apart from this, the only visible feature was a small plaque that read, "If you seek a monument, look around you." They looked around. Except for the unadorned plinth, they could see nothing but a great area of flat, featureless stone paving.

"What is there to see?" asked Dorothy in perplexity.

"I don't know. I can't see anything in particular," answered the Scarecrow. "Wait a minute, though. Look at that!" The Scarecrow, as was his wont, had observed that the plaque had changed. Now it read, "No, I mean *really* look!"

They looked again and could hardly believe what they saw. In place of the flat and even plane, they saw a region that was ludicrously *un*even. Great towering plateaus rose high in the air, irregular buttresses protruding from their sides. From these the ground descended into deep pits scattered between. Never had they beheld such a jumble of irregularity.

"How did it change so quickly?" gasped our young protagonist.

"It has not changed. Despite the change in your perception of it, it is the same pavement. You are seeing the effect of a local gauge transformation."

---

[3] Niels Bohr, Louis de Broglie, Richard Feynman, Paul Dirac, Erwin Schroedinger, and Werner Heisenberg. I apologize to everyone concerned for the random mixture of irrelevant personal characteristics and references to their specific theories.

The unexpected voice came from behind them. Everyone turned and saw two black-clad figures, one tall and slender, one petite and shapely—the Witches of Charge and of Color. No one had seen them arrive.

"This is an analogy, of course," remarked EM. "Gauge transformations refer to phase rather than to height, but because you perceive height as directly related to gravitational potential, this makes a fair analogy. I said the pavement is unchanged. The essence of a pavement is that it is something you walk on, so walk on it, my child."

Dorothy looked at the scene before her, the towering peaks and the deep chasms. Was she really expected to walk across that? However, the Witch of Charge had such a commanding presence that she found herself stepping out across the distorted landscape, though she prudently chose to start off in a direction where the ground had dropped only slightly.

She was amazed to discover that as she moved forward, some force seemed to buoy her up so that she set foot on the lower level without any jolt. This gave her confidence, so she moved forward to where the ground fell away with a drop that should have jarred her severely when she landed. Again she alighted gently, with no perceptible sensation of falling. This so impressed her that she threw caution to the winds and ran freely across the area, paying no heed to the grotesque distribution of peaks and pits. When she leapt into a deep pit, she was lowered so gently that it was as though she stepped along a flat track. When she moved toward a high peak, she soared upward and set foot upon its top with no perception of exerting any effort. She closed her eyes briefly, and it felt exactly as though she were strolling across the level surface she had observed earlier.

Opening her eyes, she saw that the others had joined her. The Scarecrow was flopping around in his usual boneless way. To look at his slack limbs, it would have seemed remarkable enough that he should move around on level ground, but now he soared and swooped from height to hollow, with no sign of undue exertion. The Tin Geek lumbered along, arms and legs flashing like pistons, as he also rose to great heights and descended to the depths. His face showed no sign that he found this in any way remarkable, but then a featureless metal cylinder can be hard to read. As for the Lion, he was bounding along with great soaring leaps from one level to another. Though he was some way off, Dorothy heard him roaring, "Remarkable experience, I must write a book about it."

As the companions gamboled about in this amazing adventure playground, Dorothy noted that although the terrain remained incredibly rough, it changed from moment to moment. Sometimes a high region would abruptly replace a trough, or a hole would fill in without warning.

Then, as time went on, she saw the ground was less convoluted. There were still humps and dips, but they were not so pronounced, and gradually they were leveling out. Soon the area looked as flat as they had seen it initially. There were no longer any unusual-seeming forces apparent, only the normal, constant force that you feel as the ground pushes up on your feet. The sense of adventure had gone, and so they returned to where the Witches were patiently waiting. As soon as they were all assembled, the Witch of Charge addressed them again.

"This interlude has illustrated for you a significant feature of any system that does not change under a *local gauge transformation*. There must be a force associated with the transformation—a force that balances the physical changes you would otherwise detect. In the case of the ground you stand on you, would not be aware of a *global* change in level. If the *entire* area rose a hundred feet into the air, it would still be flat and you could still walk around without noting any difference whatever. A local change is different. When different areas change level arbitrarily by different amounts, then you will surely notice as you walk about, unless, as in this case, there is a force present that changes in such a way as to compensate for the changes in level."

EM waved her hand to indicate the recently vanished vision of peaks and troughs.

"As I have said, this was an allegory. A gauge transformation deals not with altitude but with phase. The name *gauge transformation* is not very meaningful—merely a historical accident. It would be much better to say that a system has *local phase independence.* You may know that a change in the global phase of all amplitudes has no effect at all. The probabilities that you derive from an amplitude do not depend on the *overall* phase of all the amplitudes. A local phase transformation does not affect the relative phases of the different amplitudes. These control the interference between amplitudes and may have a great effect on the probabilities that you observe. There is no way that such gross changes could be hidden. What local phase independence means is that you cannot detect a change in the phase of *all* the amplitudes, even a change that differs from place to place. It is a statement about the amplitudes and also about space and time."

The Witch paused and bent a serious gaze upon her audience.

"The way in which the phases of the amplitudes vary from one place to another could scarcely help having some effect on what you observe. If the phases of amplitudes may be changed arbitrarily by different amounts in different places without anyone being the wiser, then this imposes tight restrictions. You would expect such arbitrary changes in phase to have observable effects, and such effects could be avoided only if they were *balanced out* in some way. This may come about if there is a force present everywhere that is affected by the changes in phase—and if that force changes in such a way as to offset those effects caused by the alteration of the amplitudes."

LOCAL GAUGE INVARIANCE

This concept is unfortunately named. It would be better called *local phase independence.*

The assumption of local phase independence is that if the phase of all amplitudes changes *locally,* then it makes no observable difference. This is certainly true for the overall phase everywhere, a so-called *global* phase change. The suggestion is that there should also be no observable difference if the phases change randomly by different amounts at different places.

Why should one even consider such an option? Well, physics advances by proposing and assessing all sorts of weird ideas. This idea would have the advantage of imposing the need for very restrictive co-operation among all the amplitudes, the sort of co-operation they would need to exhibit for the notion of renormalization to work. (See Chapter 8 for a discussion of renormalization.)

You find that such a local phase change *would* give observable effects unless there existed an **interaction** that itself changed in such a way as to compensate exactly. This interaction would have some rather odd properties, but they turned out to be the very properties that the electromagnetic interaction was already known to have. Electromagnetism is also the interaction for which the renormalization process was known to work well.

Interesting?

The Witch paused and scowled at Dorothy and her friends.

"You might ask what is the point of even considering such a complicated, absurd, and unnecessary topic," she declared. Dorothy *had* considered asking something like this, but she now thought better of it.

"The great thing about a system that *is* unchanged by a local phase transformation is that this, in itself, forces great co-operation among the phases of all the amplitudes concerned. Such co-operation is valuable, even vital. You have been told how the field around a particle contains an infinite number of amplitudes. If the superposition of these is to be controlled, even to the limited extent implied by an infinite renormalization of the charge and mass, then there must be considerable discipline and co-operation among all of the amplitudes. It turns out that sufficient co-operation is ensured if the whole system is not altered when amplitudes are changed by an arbitrary phase and a different change is made at different points. In fact, it does seem that this is the only way in which good behavior may be achieved."

The Witch paused and glared around, looking rather like an old-fashioned disciplinarian schoolmistress.

"The need to consider both amplitude and force is not so strange when you think about it. A force or interaction is, after all, just the effect of virtual bosons, and these are described by an amplitude, as are all particles. We are simply saying that when we consider the effect of changing the phase of *all* the amplitudes, we must include the amplitudes for the interaction bosons as well. This is surely only fair and reasonable.

"It is required that local changes in phase shall not affect in any way what can be physically observed. It is then the role of the 'force,' of the virtual boson amplitudes, to change in such a way as to reconcile any changes that would otherwise be observed. The notion of such co-operation should not seem too strange, given that the very idea of quantum interference is a collaborative effort between different amplitudes. In this case the requirements are particularly stringent. This need for local phase independence imposes a very specific nature—and in many ways a peculiar nature—on the force. The properties and the space–time behavior required are such that you might not find the idea plausible, *except* that an interaction is already known to have all these properties. It is electromagnetism: QED."[4]

EM stopped speaking and stood looking down on them with an air of finality. The shorter figure of the Witch of Color came forward and spoke in her turn.

"She means, of course, Quantum Electrodynamics, the theory of light and matter. That is a gauge theory, and the local phase independence defines the theory. The same is true for QCD—that is, Quantum Chromodynamics, the theory of quarks and color. This too depends on a local gauge independence, but because there is now the extra freedom given by the number of colors, the changes in phase are not just numbers but also include a rotation among the three colors.

"You can see," she asserted hopefully, "that a global rotation of the colors will not affect the binding by the color force. If red becomes green, green becomes blue, and blue becomes red, then a baryon will still contain a quark of each color and will still be bound. If a meson contained a red and an anti-red quark, then it would now have a green and an antigreen quark and still be bound. Actually, the meson amplitude would be a mixture of all three color possibilities, and that would be unchanged by the color rotation. As with the simple phase, a *local* change of color is not so easy to ignore. Random changes of color from place to place would give arbitrary changes in color similar to those that occur when a colored gluon emits another colored gluon. Such gluon emission is thus implied as a behavior of the compensating force.

The need to compensate and suppress all overt signs of such behavior puts very tight restrictions on the form of the interaction and forces good cooperative behavior amongst the amplitudes. It gets a bit complicated," she admitted finally. "There are a few obvious features, though, such as the fact that the interaction bosons should be massless."

"Why is that then?" asked Dorothy. She had little doubt that the Witch planned to give some sort of explanation, but she noticed that the Lion was sidling up with a book titled *He Ain't Heavy, He's My Boson* by Isa C. Lion. Fortunately, the Witch spoke again before he could reach her.

---

[4] Here the Witch is making a play on words (or letters, rather). Q.E.D. is the abbreviation that is used at the end of mathematical proofs and that means, more or less, "I told you so."

"In its role as a so-called gauge field, the interaction has a mission of conciliation. It must be able to compensate for any observable effects that would otherwise appear, and to do this it must be present everywhere. It must be long-range, with lines of force extending out to infinity like those of the electromagnetic field.[5] This means its virtual bosons must be massless."

"Does that follow?" queried our heroine. She was becoming rather tired of such casual assertions.

"Oh, yes. Most certainly. You know that a field boson is emitted by some particle that has no energy of its own to spare for particle creation and must borrow energy that it does not have. It is thus a virtual particle, and any energy it has must be borrowed, as laid out in the terms of the time–energy relation. If it needs energy to create rest mass, then it cannot have it very long and it cannot get very far. Massive exchange bosons give *short-range* forces, and the flux lines peter out close to their source. If the lines are to extend out to infinity, if the force is to be long-range, then the bosons must be massless. If there is no mass, there is no need for energy to create it, no lower limit on the energy that the virtual bosons require, and so no upper limit to their range. The force will have infinite range, falling off in strength with distance only because the same flux is spread over a wider and wider area.

"That is enough for now," concluded the Witch of Color. "I am sure you have heard all you want to hear about gauge theories. Possibly a bit more, even. You should be on your way now. There is more to see, and you will not find it among these monuments. You have still to meet the last of our sisters, the Weak Witch."

"She does not sound very significant," remarked Dorothy.

"You cannot judge from a name, or even from an interaction rate. She is a strange one, and she can do things that we cannot. She *changes* quarks. Continue across the Plaza of the Immortals and out of the city, and I am sure that you will soon find her, or she find you. So move it, girl."

So instructed, Dorothy and her companions followed the directions, and soon they were out of the city and walking along a quiet country lane.

---

[5] This is not perhaps a totally convincing argument for massless "gauge bosons." There are other arguments that *are* convincing, but they depend on the transformation properties of the mass term in the field Lagrangian, on relativistic properties of the particle spins, or on the Bohm–Aharanov effect. Believe me, you don't want to know about them.

CHAPTER 11

A
WEAK
OLD
WOMAN

As they walked along the lane, Dorothy spotted some strange-looking flowers in the hedgerow.

"Those look strange," she exclaimed.

"Yes, they *are* strange. They are Kaonemonies and they contain kaons, which in turn contain strange quarks," commented the Scarecrow, who had observed them before.

"What are strange quarks?"

"Well, obviously they are quarks with strangeness."

"And what is strangeness?" asked Dorothy, though with no great confidence of receiving an answer. Sure enough, before the Scarecrow or anyone else could respond, they all noticed that the flowers were *changing*. Not all at once, not suddenly, but gradually, the vision of the exotic flowers was overlaid with a much more common-looking blossom, rather like a daisy. The strange blooms slowly faded, and there remained only mundane buds, indistinguishable from others around.

"It's her!" called the Scarecrow, heedless of grammar. "Her! The Weak Witch. She must have done this. And she must be nearby, for hers is a short-range interaction."

They all looked around, and Dorothy saw, slinking away through the hedgerow, a short figure totally muffled in a hooded black robe. Quickly they

R Gilmore

gave chase and followed the half-glimpsed and elusive figure through the trees. They lost sight of her just before they came to a clearing in the wood. In the middle was a little fairytale cottage. There was no sign of the black-clad figure they had been following, but it seemed probable that she was in the house, so they marched up to it and the Lion confidently rang the door-bell. After a brief delay the door was opened by a figure in black, a hood pulled up over her head so that her face was completely lost in shadow.

"Hello, my pretties," she cackled in an appropriately Witch-like voice. "Are you lost in the woods then, you poor little dears?"

"No, we are not," boomed the Lion confidently *and* assertively. "We are only about a hundred yards from the road and we are not little dears either. You are the Weak Witch, aren't you? Dorothy here wants to know all that you can tell her about the weak interaction." Dorothy wasn't so sure about that, but she let it pass.

"OK, OK," snapped the Witch, reverting to a normal voice. "If you won't get into the spirit of the thing, you had better just come in."

She led them through her cottage and into a small kitchen at the back. The entire dwelling had a strangely appetizing feel to it, and a large black pot bubbled over a fire at the far end. The Scarecrow commented on the fla-vorful atmosphere of the house.

"Well, I do have a way with flavor," admitted the Witch rather smugly. "Pride of place goes to my flavor shelves."

She indicated six little sets of shelves on one wall of the narrow room. These were arranged in two groups, each comprising three sets of two shelves, one above the other. On the opposite wall there were more shelves, apparently identical to those on the first wall.

"The three on the left are for leptons, the three on the right for quarks," explained the Witch. "In the quark shelves I have different flavors of quarks on the top and bottom rows. On the top shelves I have the 'up', the 'charm,' and the 'top' flavors of quarks. On the level below there are 'down', 'strange,' and 'bottom' quarks. This has the effect of grouping the quarks according to elec-tric charge. They are arranged in doublets, one in the upper level being paired in each case with one in the level below. The quarks on the upper shelves have greater electric charge; those on the lower shelves have less."

"I see. Does that mean that they have a single unit of charge on the top shelf and none on the shelf below, or perhaps two on the top shelf and one below?" asked the girl. She liked to take opportunities to air her new knowl-edge. She had been told on her visit to Cern that the electric charge carried by a particle was always some multiple of the charge on the electron, al-though charges might be positive just as well as negative.

"No, not in the case of quarks. The quarks on the top shelf have one unit of charge more than those on the lower level, it is true, but none of them has a charge that is a whole number of electron charge units. The higher ones have a charge of $+\frac{2}{3}$ of a unit, the lower $-\frac{1}{3}$."

"But that can't be so!" protested Dorothy. "I was clearly told in the King-dom of Cern that every particle they observed either had no charge or, if it had a charge, then it was a whole number of electron charges." The Scare-crow nodded so enthusiastically at the reference to observed charges that it looked almost as though his head might fall off. Dorothy suspected that in his case this might be a very real danger.

"Perhaps that is so for observed particles," retorted the Witch, "but quarks are *never* observed. You can never release a free quark, so you do not observe the electric charge of one on its own. Because it takes three quarks to make a baryon, and because the baryons have charges that are multiples of the standard unit, it follows that the quarks will have charges that are mul-tiples of *one-third* of the standard unit. And that is what the charges are found to be. Some are two-thirds and some are one-third, but none has a charge that is a whole number of standard electron units.

"If you want integral charges, then you must look to the leptons," she continued, pointing to the other three sets of shelves. "They are not confined, they are free, they are out in the open for all to see, and *they* do have charges that are a whole number of electron charges. In the first set you have the electron. It has one single electron charge, not too surprisingly. The other two sets hold the other 'generations' of leptons: the muon and the tau. On the lower shelves in each case are the appropriate types of neutrinos that are paired with each of the three charged leptons. You may not have met neutrinos before. They are my own, my very precious darlings."

LEPTONS AND QUARKS

There are three "families" of both leptons and quarks. Each family contains a "doublet" of particles. The quarks also come in three "colors"—red, green, and blue—as discussed earlier.

Leptons			Quarks		
e	μ	τ	u	c	t
ν_e	ν_μ	ν_τ	d	s	b

In the case of the leptons, those in the top row have the same charge as the electron, whereas the various *neutrinos* ν in the second row have no electric charge at all. The quarks on the top row have a charge that is $+\frac{2}{3}$ of the standard charge, and the quarks in the lower row have $-\frac{1}{3}$ of that charge.

"I see nothing."

"Neither do I."

They all glanced around to see who had spoken. Cramped into the remaining space in the narrow kitchen they saw EM and the Witch of Color. No one had seen them arrive (but then of course they had been there all the time).

"I doubt that you do, my dears," crowed the Weak Witch. "There is nothing for either of *you* to see. The neutrinos are there all right. There are three different varieties of them no less, but they have no electric charge, so you do not see them." She turned toward EM as she said this. "Leptons in general have no color charge, so *you* do not see them either," she continued, nodding her cowled head toward the Witch of Color. "The neutrinos are all mine, for me alone to see. Because they escape the attention of the dominant interactions, they are slippery, sneaky little creatures that can go an awfully long way without anyone being aware of them. I love them," she added.

"Neutrinos pour out of the Sun," she went on, addressing Dorothy directly. "A great multitude of neutrinos fall upon the Earth, as do the photons that constitute the light from the Sun. The difference is that nearly all of the neutrinos pass straight through the Earth without even being aware that it is in their path." She looked fondly at the bottom row of shelves, in each of which Dorothy thought she could perhaps make out the faintest ghost of an apparition.

"The first generation contains the workers. Among the leptons this means the electron. Amongst the quarks the first generation contains the up and down quarks—the quarks from which are made both the proton and the neutron."

"Why are there two other generations then? Dorothy whispered to the Scarecrow, who was nearby. "I don't know. Perhaps for company?" whispered the Scarecrow in response.[1]

"You have met all these particles before," continued the Witch. "Apart from the neutrino, of course. But where my weak interaction differs from all the others is that I have the knack of turning one flavor into another. The universe is full of protons and electrons in the form of hydrogen. Why isn't the universe full of neutrons also?"

"Isn't it?" Dorothy whispered to the Lion. "No," the Lion roared back quietly.

"It is because the neutron *decays*. A neutron is heavier than the combined masses of a proton and an electron, and so there is some energy to spare, energy that can be released in a decay process that gives these two particles. In fact you end up with three particles, as I shall explain. The 'down' flavor quarks can decay to 'up' flavor quarks, and it is my weak interaction that causes this decay. The virtual bosons of the weak interaction can carry charge, and when such a boson is emitted or absorbed, the particle that emits or absorbs it will change its electric charge and transfer between the upper and lower positions in a group. A neutron contains two 'down' quarks and one 'up' quark. One of these 'down' quarks can emit a negatively

---

[1] I don't know either. As far as I am aware, nobody knows why there are three generations of leptons and quarks, the extra two sets roughly mimicking the properties of the first. Interestingly, there do not seem to be more than three sets. There is good evidence that there are only three types of neutrinos. If there were more types of neutrinos, then the Z boson (see later) could and would decay into them as well as into the known types and so would decay more quickly than is observed.

charged boson, and in the process the 'down' quark must increase its charge correspondingly and so become an 'up' quark. This leaves a baryon that now contains two 'up' quarks and one 'down' quark. The neutron has become a proton. The negatively charged virtual boson may then be absorbed by a neutrino to produce a negatively charged electron—and there you are," finished the Witch.

"Does that mean this weak decay can take place only if there is a neutrino around to absorb the boson?" asked Dorothy. This didn't sound too unreasonable after what the Witch had said about the huge number of neutrinos passing through the Earth.

"Well no, not really. And I mean that in the technical sense," replied the Witch. "There is no need for a *real* neutrino. The boson may be absorbed by a virtual neutrino in a neutrino–antineutrino pair that has been created in the vacuum. Remember that the vacuum is full of particle–antiparticle pairs, and this will include neutrinos. Because there are so many of them, these will be much more readily available than any so-called *real* neutrinos, in the same way that the intensity of virtual photons may be so much greater than that for real photons in any conceivable beam of light. This virtual neutrino is converted to an electron, and its companion virtual antineutrino receives enough energy to become real. The neutrino being virtually massless, this doesn't actually require very much energy. If you add everything up, you will see that the observable effect is that an initial neutron has gone, decayed. In its place there is now a proton, an electron, and an antineutrino—three particles in all. This process, when it occurs to a neutron within a nucleus, is called β decay."

"I asked the Nuclear Inspector about β decay," remembered Dorothy, "but he didn't answer."

"Perhaps you now realize why he chose not to," observed the Scarecrow.

"My bosons can only cause a transfer between the upper and lower members of the same doublet, " continued the Witch, "but that is all that is needed for β decay, because the 'up' and 'down' quarks *are* in the same doublet."

"But then it doesn't explain the decay of the strange quarks in kaons," mused Dorothy, who had been attending rather carefully and remembered what the Scarecrow had said just before they observed the Weak Witch. "From all that you say, the 'up' and 'down' quarks are in a different doublet from the 'strange' ones, and you say also that your virtual bosons can produce a change only between two members of the same pair." She pointed to the doublet of shelves that contained the 'up' and 'down' quarks in order to emphasize her point.

"Just a minute there, people," interrupted the Witch of Color. "That's not a 'down' quark in the lower shelf. Not purely. That's a mixed state, a superposition of quarks," she added to be quite clear. Dorothy looked at what she had been taking as a 'down' quark. She couldn't be at all sure, but perhaps it *did* look a bit mixed.

"Nonsense," replied the Weak Witch in response to the Witch of Color. "That is one of my flavor states sure enough. It is the sole companion for the 'up' quark, pure and simple. There is no doubt about it." She leaned over to peer more closely at the contentious state and swept back her muffling hood in order to see more clearly. As her face emerged from the deep shadow that had hitherto concealed it, they saw that she bore a remarkable resemblance to EM. She was shorter and more heavily built, but otherwise the family resemblance was remarkable. Her Witch-like appearance was further enhanced by the fact that she was cross-eyed; her eyes glared in quite different directions.

"There is *every* doubt," responded the Witch of Color aggressively. "You may be half blind, but I can see a quark well enough and know whether I am seeing a single pure quark state or a mixture. That is a mixture," she finished firmly.

The two Witches had moved closer together and were now face to face. Dorothy was afraid that the situation was about to get ugly when an unexpected peacemaker joined the fray. It was the Confident Lion.

"It is quite simple, really," he began. It was not perhaps the most tactful of openings, but it did distract the two Witches. They were further distracted by his reaching behind a saucepan and producing a book titled *The Working Weak* by Isa C. Lion. He unhesitatingly opened the book at a page that Dorothy noticed was marked by a bookmark. "These bookmarks are quite invaluable," he muttered. "Here is the section I need. Right here under the title 'Weak Eyesight.' You are both right, ladies. Each from your own point of view. For the strong interaction, which means for the color forces, 'down' and 'strange' quarks are distinct and different. You can quite clearly tell them apart," he remarked, nodding his head with its great mane toward the Witch of Color. "A 'strange' quark has strangeness. Though this is no more than a name that is given to whatever characteristic makes a 'strange' quark distinctive and different from a 'down' quark, you are clearly aware of that distinction. They both have the same electric charge, one-third of the charge carried by an electron, but they are not the same as each other. The strong interaction cannot casually change a 'strange' quark to a 'down' quark. They are different, and as far as the strong interaction is concerned, they remain so. In your hands 'strangeness' is conserved.

The weak interaction, on the other hand," he said, turning his head to the Weak Witch, takes no account of 'strangeness.' You do not see it at all. For you the 'down' and 'strange' quarks are much the same. You do not distinguish them, and the states you see might just as readily be a sum of both—so consequently they are. You do see a quark state in the lower position in a doublet, and this state is paired with the 'up' state, but the state that you see is not quite the same as any of the states seen by the Witch of Color. You see things differently and have different criteria. The state from which your virtual charged bosons can produce an 'up' quark is not exactly a 'down' quark. In fact, it is a mixture of 'down' and 'strange.' Both 'down' and 'strange' quarks can decay to an 'up' quark because both types are present in the specific amplitude that you and your interaction see as paired with the 'up' quark. The state contains a smaller amplitude for 'strange' than for 'down' flavor, so the 'strange' quark decays more slowly, that's all."

Dorothy felt that was quite as much as she wanted to know about strangeness and the transformation of strange quarks, so she decided a change of subject was in order. She commented on the Weak Witch's remarkable resemblance to EM, the Witch of Charge.

"Yes, we are sisters," replied the Weak Witch. "Though we seem very different, we are in many ways the same. Close twins of each other. Our interactions have the same strength, for example."

"How can that be?" exclaimed the girl in astonishment. "I have been told again and again that EM's strength and reach are impressive, whereas your interaction is, well, weak."

"I said that our strength is the same, not our reach. The range of my weak interaction is short, and so, although the strength of the coupling to a particle may be as great as for electromagnetism, the rates of decay produced are much less." She paused as though all were clear.

THE ELECTROWEAK INTERACTION

Nowadays the electromagnetic and the weak interactions are considered to be *unified* into a single electroweak interaction, despite their apparently great differences.

They are felt to be essentially the same because the strength with which the interaction bosons couple to their source is much the same throughout. These bosons are the photon, the W^+, the W^-, and the Z.

The photon is massless, and it has no charge itself, but is emitted only by electric charge. There is no discernible change in a particle after it has emitted a photon.

The other three bosons are very massive, about a hundred times as massive as a proton. This gives the interaction an extremely short range, and it has much less effect in causing transitions and decays than the long-range interaction carried by the massless photon. These bosons do not couple to electric charge specifically but to some generalized weak charge—they couple to uncharged neutrinos, for example.

The Z has no charge and leaves the particle that emitted it unchanged, much as though it were a heavy photon.

The two W bosons carry off electric charge and alter the particles that emit or absorb them. They can change quarks and are responsible for the decay of strange quarks and for nuclear β decay.

"I am afraid I do not quite follow that," said Dorothy as politely as she could.

"Perhaps I can make it clearer." The interruption came from the usually silent Tin Geek, who clanked forward with his chest screen beginning to glow. "You must look at the interacting particles. Each will have an amplitude spread over some region." On his display screen appeared two blurred, overlapping regions that represented the particle amplitudes. One region was shaded in pale grey; the other was darker. This was solely to help distinguish the two.

"Look at an interaction between the two particles. If the interaction has a long range, then each point on one amplitude will be able to interact with every part of the other." In the picture displayed on his chest, a set of expanding rings appeared from a point in the dark amplitude and spread across the entire screen, totally encompassing the paler amplitude. A second set of rings appeared from another point on the dark distribution, and they too spread over the entire pale region. Again and again, such expanding circles appeared from different points on the paler amplitude, and in each case they covered every part of the other.

"You see that in this case one particle is totally connected to the other. Every part of the first amplitude is aware of every part of the other through the interaction. It is different if the interaction range is short."

Again the two blurred regions appeared, and again a sequence of expanding rings spread out from different points on the dark distribution. This time, however, the rings faded and vanished almost before they had left their point of origin. In each case they spread over the tiniest fraction of the full extent of the pale particle amplitude. The dark distribution was connected to the pale one only in the immediate vicinity of each point of overlap.

"That is the short-range case. One amplitude connects with another only where they overlap. With a long-range interaction every point reaches every other. When the range is short there is much less interaction, even if the exchange bosons couple as strongly to the parts of an amplitude that they do reach, because they can reach so much less of it.

As particle energies and momenta increase, the particles become more localized. They are more compact. A short-range interaction can cover more of these compact amplitudes. When particle energies increase, interaction probabilities for short-range and long-range interactions are not so different."

"That is true," agreed the Weak Witch. "My exchange bosons interact just as strongly as EM's photons, but they are heavy. They have a large mass, so a great deal more energy is needed to create them virtually. So great an energy is available for only a short period, even from our forgiving Universe. The bosons cannot get far before their energy loan must be returned, and therefore the range of the interaction is short. All of this is borne out by observation. Observe!"

With this abrupt command the Witch raised her hood to cover her face, pushed up the sleeves of her robe, and whirled around to face the fire. She threw something into her bubbling pot, which looked more like a cauldron than ever. A dense column of smoke rose from the liquid within and curled twisting through the air to settle upon the surface of the table. There it thickened and condensed to reveal a small creature that, upon realizing its predicament, shied in alarm with a clatter of tiny hooves. The new arrival looked like a diminutive horse with a neatly suited doll attached to the front. It was the Information Centaur, shrunk to a minuscule size.

"Oh no, not again!" he cried, looking in alarm at the great figures towering over him.

"I have summoned you before us, for we have need of information from the land of Cern," intoned the Witch formally, completely ignoring his reaction.

"Why do you keep doing this to me?" moaned the Centaur. "Why can't you use the Internet like everyone else?"

"Be silent! Tell us what you can about the scattering of neutrinos and the nature of intermediate bosons."

The Centaur correctly interpreted this self-contradictory command and decided that the sooner he complied, the sooner he would be able to return to his normal occupations.

"OK, what do you want to know?" he said resignedly.

R Gilmore

"Show us the weak interaction when it is no longer weak. Show us the scattering of neutrinos at high energy."

"Very well," replied the Centaur. He sketched a rectangle in the air and it became a window—a window in space that rose and grew to show a view of an experimental hall like the one they had seen at Cern. In the center of the hall was a massive apparatus much like the one Dorothy had seen installed within the accelerator tunnel. The Centaur explained that this was a neutrino interaction experiment. Even though the neutrino amplitudes had become more compact at the high energies involved, this gave them no great interaction with bulk matter compared to the long-range electrical interaction of a charged particle. The beam of neutrinos could pass readily through the shielding material around the accelerator and into the detector. There it might interact with individual nuclei as might a beam of electrons.

The window now showed a picture of the detector's interior, much like the view that Dorothy had seen on a screen when she was herself at Cern. The only striking difference was that previously there had been a multitude of tracks across the chamber as charged particles passed through without any major interaction. There were no such tracks now, because the neutrinos did not have the long-range interactions that registered in the material of the detector. The volume seemed totally empty, and then suddenly a great jet of particles showered forward, as dramatic as any they had seen before. The jet seemed to come from nowhere. There was no sign of any incoming particle on the left, for the neutrino left no trace. Then, from no source that was apparent, there proceeded this great spray of particles, all moving toward the right with grotesque imbalance.

"That is a neutrino event," said the Centaur. "High-energy neutrinos will interact quite readily."

"But the real evidence for a short-range interaction is the release of the exchange bosons as real particles and the observation that they are indeed heavy," remarked the Weak Witch. "The masses of the bosons necessary to give the observed β decay rates can be calculated if it is assumed that the interaction is basically as strong as electromagnetism. These are the masses that were observed. Tell them!" she commanded the Centaur.

"It is true," he began. "Collisions between a beam of protons and a beam of antiprotons have created such bosons. The protons and antiprotons are made from quarks and antiquarks, respectively. The Witch has told you how a 'down' quark may emit a virtual W^- boson, a negative exchange particle, and leave behind an 'up' quark. You may see that a collision between a 'down' quark and an anti 'up' quark can leave such a boson on its own. If the colliding particles have sufficient energy, the resultant boson may be given enough energy to produce its rest mass and be released into the world as a real particle. Both positively and negatively charged W bosons were found, and they were very heavy. They had masses about one hundred times that of a proton. Such a huge mass for its virtual exchange bosons gives the weak interaction a *very* short range, and such a tiny range for the interaction results in an apparently weak β decay. You need me no longer," he concluded. "The results were fully reported in this paper."

Another swirl of smoke came coiling and twisting from out of the Weak

Witch's pot and obscured the tiny figure of the Centaur. When the cloud cleared, the Centaur was gone and a technical publication lay upon the Witch's table. Dorothy bent over to look at it, but she could make nothing of it. It was far too technical.

"Of course," remarked the Weak Witch, "you may ask why there are neutrons in atomic nuclei." Dorothy had not thought of asking this, but now that the Witch mentioned it, she supposed that she might have. "The universe is filled with hydrogen, which is to say with protons and electrons, but the free neutrons that were originally present have decayed. If there were no neutrons readily available to build into the nuclei, then where have the neutrons that *are* in nuclei come from? You might well ask that," she repeated, "and I shall answer you. First, though, you should all go outside and go around the back of my cottage. When you get there, you must inspect my garden of SUNflowers."

Though they could see no reason for this excursion, Dorothy and her companions had learned better than to argue with the Witches. They obediently trouped outside, and as they left, they saw the Witches cluster about the bubbling pot. "When shall we three meet again?" shrieked the Weak Witch with a return to her weird persona. "In about ten minutes, I should think," responded EM unsympathetically.

The companions made their way to the back of the quaint cottage. There they found that there was a great clearing in the woods and the Witch's SUNflowers were dotted around this dark field. The earth appeared black, though on closer examination they were not so sure that it was earth at all. The spaces between the widespread flowers seemed to be just that: black, empty space. They looked carefully at one of the nearest flowers. It had a round face, not a dull yellow like a normal sunflower, but an intense blaze of blinding light. Around this brilliant central disk spread wide coronal petals.

The group made their way toward it, and as they approached, it was as though the flower moved steadily from them. It grew larger as it retreated, and their approach became faster in response, until, at the end, they found that they were in close orbit around the furnace surface of a fiery star. As they floated along, they found themselves approaching another group of figures. There were three, dressed in garments as black as the space behind them and apparently floating in front of a dark captive balloon. After a moment, they realized that the balloon was in fact the Witch of Mass and that the three figures were the other Witches.

"Here you will witness the birth of atoms. Stars are the cradles for the nuclei around which form the atoms of your world," asserted the Weak Witch. "Every one of us plays her part in this stellar alchemy."

NUCLEOSYNTHESIS:
THE BIRTH OF ELEMENTS

After the rapid events of the Big Bang, the matter in the Universe was mostly protons and electrons, combined to give hydrogen and also a little helium and less
deuterium. There was no carbon or oxygen or any of the other heavy elements that
are necessary for our life. These were all forged in the furnace centers of stars.

As noted below, the various interactions all play a part in this.

- Gravity assembles the matter and releases enough potential energy, as the matter falls, to heat the center of the star hot enough for nuclear processes.
- The repulsive electrical interaction keeps the nuclei apart, so that they can penetrate the "electrical barrier" only if they have very high energy. Even then, reactions are improbable and the process is slow. Stars last a long time.

- The strong nuclear force sticks the lighter nuclei together to form heavier ones, but there are too few neutrons to make stable nuclei.
- The weak interaction may change some protons into neutrons. This provides the constituents to make stable heavier nuclei.

The details of the processes are complex, but by one means or another the nuclear cores of our elements are made within a star. They are of course still in the center of stars, but in their death throes, stars expel material back into space, and after a long time this material may be used to form a planet, like ours.

This is what people mean when they say "we are stardust," though perhaps we should say "ashes from a dead stellar fire" instead.

"My role comes first," began the Witch of Mass. "The stellar play begins with a space filled, if rather sparsely, with hydrogen gas: protons and electrons that have lasted since the founding of the Universe and have had plenty of time to pair up with one another. In some regions the gas may be a little denser, a little thicker, than in others. Every atom of gas has a tiny gravitational interaction with every other, and where there is more gas present, it exerts a greater pull. It pulls in gas from other regions and grows denser still, becoming yet more attractive in the process. This process continues at an ever-increasing rate until you have the early stages of a star."

The Witch waved a massive arm, and a window opened against the dark background of space. Initially the window was not much different from this background. It showed a cloud of gas forming in space, with more and more wisps drafting in and moving faster and faster as they approached.

"As gas is drawn down by the gravity of this protostar—this celestial starlet—it gains energy. Just as a cup dropped from a height above the surface of the Earth will gain energy it moves faster and ever faster. When the cup hits the ground, this energy must be converted. In the case of the cup, the energy goes into breaking the bonds of the cup's substance and making the sound of a crash. When the gas falling into a star collides with the dense center, its energy is converted to heat. The particles within the star move faster and collide with appreciable energy. Initially, the energy in the collisions is not very great, and the collisions have little effect beyond the emission of photons. The protostar begins to glow with its own light."

In the conjured window, the mass of gas indeed had begun to glow. Not as yet with the full brilliance of the Sun, but its intensity rose dramatically as more and more material contributed to the energy.

"As the star grows and its mass increases further, so does the energy gained by the incoming hydrogen. The temperature rises, and this means that the energies of the particles are greater. Eventually they are great enough to produce nuclear interactions."

"My role in this is to provide a steadying hand. I keep the reactions within reasonable bounds. I keep the star's behavior sane and measured," put in the Witch of Charge. "All of the protons carry a positive charge, a charge that is concentrated in a tiny region. When they get close to one another, they are repelled by their like charges and so avoid an intimate collision that might give a nuclear interaction. Only at the highest temperatures will a few—a very few—protons have energy enough to overcome the barrier of this repulsion and have an encounter sufficiently close for their nuclear fields to engage. Without such inhibition, a star might burn out rather quickly, its life best measured in minutes rather than in millions of years. That could be inconvenient," she finished dryly.

"Where the protons do make close contact, the strong nuclear forces can engage." The Witch of Color took up the tale. "But such is the contest between the repulsion of the electric charges on the protons and the residual attraction that may leak out from the color forces within them that no stable objects can be made from protons alone. You must have neutrons also to balance the states. But there are no neutrons available."

"Or there would not be save for my so-called weak interactions," broke in the Weak Witch. "If there were neutrons available, then bound states could be formed by strong interactions. Without neutrons they cannot. If such interactions were to take place, then energy would be released, and it is part of the basic thermodynamics of nature that it much favors the release of any potential energy to provide the kinetic energy of moving particles. It is such energy that gives you sunlight.

"The weak interaction saves the day. Out in free space the neutron has more energy tied up in mass than do the proton and electron that it could produce, and so it is normally neutrons rather than protons that decay. Within the intense furnace of a star, the presence of neutrons will allow the formation of bound states and, in the process, release more than enough energy to make up for the extra mass of the neutron. In such special circumstances, the weak decay runs in reverse and the protons can become neutrons. They may absorb an electron and emit a neutrino along with the resultant neutron, or they may emit a positron, an antielectron. Because this entity would annihilate with an electron already present in the star, it would come to the same thing in the long run. And there you have it, the chain of interactions that links a universe of hydrogen to a stellar core that is packed with the nuclei needed to form the atoms of which your world is composed. All that is needed then is an explosion to spew the nuclei out into space to give the seeds of future worlds. But that is another story."[2]

Dorothy and her companions could hear a saccharine cosmic choir singing

> *When you wish upon a star,*
> *You get the atoms that you are.*

They looked around and now saw no sign of the Witches, only the impersonal glare of the nearby star's photosphere and the scattered specks of other distant stars. They found themselves moving together away from the star, and as they moved, the distant specks of light seemed to group together, more like candles in the night or like the lights of a remote town. Then they became less remote, but at the same time much less clear, and only the closest could be made out, shining wanly as if through dense smog. They could feel ground beneath their feet again and realized that they were walking over cobbles, along a street that was dimly lit by occasional gas lamps that battled unsuccessfully with the surrounding smog. On a wall nearby they could just make out a metal plate that read *Candlestick Street.*

---

[2] This is a story that I hope to tell, among others, in a future book. It is tentatively to be entitled *Once Upon a Universe*.

CHAPTER 12

THE

HIGGS

OF THE

MASSKERVILLES

Dorothy and her companions walked on down the smog-choked street until the sound of a door opening nearby attracted their attention. The brass plate by the door announced that this was number 221A Candlestick Street, and the open doorway revealed a woman in a housekeeper's severe dress and apron.

"Please come in, lady and gentlemen," she accosted them, ignoring the fact that Dorothy's companions did not fit well into the latter category. "Mr. Ghardens has been expecting you."

They all accompanied her inside and up a flight of stairs. They followed her willingly both because their curiosity was excited and because they were glad to be out of the choking smog. Their guide threw open a door at the top of the stairs, stood aside, and announced, "The lady and gentlemen you were expecting, Mr. Ghardens."

The room was furnished with a certain bachelor casualness, not to say eccentricity, and contained two people. One was a solid, rather military-looking man, dressed respectably in a frock coat and sitting stiffly in an upright chair. The other was the more striking of the two, with fine aquiline features and a strong jaw. He was reclining languidly on a sofa, wearing a purple silk dressing gown somewhat stained with chemicals. He was idly leafing through a newspaper while smoking a noxious meerschaum pipe. When he saw them, he rose to his feet.

"Ah, come in, come in," he greeted the newcomers exuberantly. "I am Sherlock Ghardens, at your service. And this is my friend Dr. Jouleson," he added, introducing the other occupant of the room. "I have been expecting you. I see that you have but recently come from the Emerald (and Ruby and Sapphire) City, where you spent some  time at the Gauge Theory monument and then came here by the way of the Weak Witch's garden of SUNflowers."

"Really, Ghardens!" exploded his companion. "This is too much. How can you possibly know such things."

Ghardens resumed his seat and began to puff complacently on his pipe. "Why, it is but a commonplace to anyone who is capable of the least deduction from what their eyes show them. You can scarcely escape observing the slight residual flickering of red, green, and blue colors about them, a lingering sign of their exposure to a demonstration of color forces. This implies a visit to the city and probably an encounter with the Witch of Color. You will note also the unconscious care that they all take when walking, as if they doubted the reliability of level ground. Such circumspection is invariably occasioned by a visit to the Gauge Monument."

"Amazing!" exclaimed Dr. Jouleson. "Your reasoning never ceases to surprise me. But how can you maintain that they visited the Weak Witch's garden?"

"Really, such things are self-evident. The weather has been overcast of late, as indeed it almost invariably is in these parts, yet you will note that this young lady is noticeably bronzed by the Sun and even copper-colored in places from the effects of the stars. She could readily have encountered such radiation only in the SUNflower garden. It is almost too trivial to mention.

"But enough of such commonplaces," he continued. "We must deal with the purpose of these good people's visit."

Dorothy had been wondering what *was* the purpose of her visit. She said so.

Ghardens raised a fine eyebrow in mild surprise. "Why, surely I need not draw your attention to the remarkable behavior of the Weak Witch's massless bosons," he said.

"But they are not massless!" objected Dorothy.

"That is what is so remarkable."

"How do you know about all this, Ghardens?" exclaimed the good doctor, unable to contain his astonishment. "All these little what-do-you-call-thems. It is astounding.

Ghardens smiled fondly at his friend. "Elementary particles, my dear Jouleson," he replied succinctly. "I try to keep up with the field. You never know when such information may come in handy."

"You surprise me. I remember that when I told you that the Earth moves around the Sun, you said that you would endeavor to forget the fact as soon as possible."[1]

"Well, I have changed my mind," snapped Ghardens. "I have come to realize that there is no absolute limit to the information that the diligent student may remember. Anyhow, Jouleson, we should be concentrating on the Matter of the Mysterious Missing Masses, as you would probably term it in one of your overly dramatic accounts. We know," he continued, leaning back and bringing his fingers together in a steeple, "that we are dealing with bosons that are not permitted to have any mass and yet have been observed to have a considerable amount. This seems a pretty problem. Now some particles are born with mass; some may, who knows, achieve mass; and some have mass thrust upon them. I think we shall find this last to be the case."

Ghardens rose, went over to a set of shelves in the corner of the room, and rummaged among the papers there. "I have a monograph on the subject somewhere here. It was written by a Professor Higgs, of Edinburgh I believe. In it he proposes the seeds of an interesting notion—that particles that provide the bosons of an interaction may have no mass of their own, but may be given mass by other particles. These particles have consequently come to be known as Higgs bosons, after the good professor."

He abandoned his search and returned to his seat. Once seated, he swiveled to regard his guests with clear, penetrating eyes. "It is a capital error to theorize without facts. Before we continue, let us consider some of the aspects of particles with and without mass. Mass is equivalent to inertia. Let us say that you accelerate a massive particle. In the process you give it more and more momentum. Initially it will move slowly and then will steadily build up speed until eventually its speed approaches the velocity of light. Then the speed will level off, though the momentum and energy will continue to rise. The lighter the particle, the more readily it will accelerate. At the extreme of the range, you have the behavior of a particle with no mass at all."

Ghardens leaned back and closed his eyes, but continued to address his audience. "If a massless particle is given any energy or momentum at all, it

---

[1] In *A Study in Scarlet*, Ghardens is portrayed making this remark, under the transparent pseudonym of Holmes.

will immediately travel at the velocity of light. That is the only speed at which a massless particle *can* travel. Photons have no mass. That is why light travels at the 'speed of light.' That speed has no specific reference to light as such. It is the natural and only velocity for any particle that has no rest mass. That is the situation for real particles," he continued, opening his eyes and again fixing those assembled with his piercing stare. "For a virtual particle that has no inherent mass, there is no particular limit on its range. At the extreme, because it needs no energy to create its nonexistent rest mass, it may borrow its energy for as long as it likes and go as far as it likes. As you know, the range of massive virtual bosons is limited because they must return their borrowed energy, but a massless boson need not actually borrow anything. The range of the photon interaction is thus infinite. It falls off in strength the farther you look from its source, but this is because it is spread over a greater and greater area.

"So, to sum up," he stated, putting his fingers together once more in a judicial pose, "photons are massless particles. Real photons travel at the velocity of light, because they are light, and virtual photons give an interaction of infinite range, and yet. . . ." He paused significantly. "There are circumstances in which light travels more slowly than "the velocity of light," and there are situations where the range of photon interactions, of an electric field, is short—much like that for a massive exchange boson."

He paused and began to stuff tobacco into his pipe. This was obviously a deliberate move to increase the tension, but the Lion seized upon it as an opportunity to speak.

"I imagine that you are going to tell us . . . ," he began.

"The parallel is clear," cut in Ghardens incisively. Dorothy was impressed to discover someone capable of cutting the Lion short in mid-sentence. Ghardens reached down to the floor beside him and picked up a large magnifying glass. Instead of using it to inspect anything, he held it out for them all to observe. "This useful device has a lens that is curved, thick in the center, thinner towards the edges. Light passing through the glass moves more slowly than light that passes through air. Consequently, light that passes through the center arrives later than light that passes through the outer parts of the lens and gets out of step. The resultant interference gives a distorted view of whatever is on the far side of the glass—and in this case a magnified one. Why, you might ask, does the light pass more slowly through the glass? The light is still composed of photons after all, and the photons are massless, are they not? It is because the photons interact with the electric charges within the atoms of the glass, and the overall effect is to confuse and hinder the passage of the light. Though photons on their own can travel only at the speed of light, in this case they are held back by the press of the crowd of charge that surrounds them.

"It is again the effect of other charges that may work to limit the range of the electrical force. An electric field normally has an infinite range, but will fall away quickly inside an electrical conductor. This is because there are electrons within the conductor and they are free to move. If there were an electric field within the conductor, then it would provide a force on these electrons and they would move. As such charges move, they change the electric field, because charges in different positions give different fields. The electrons move until the new field that they have made cancels out the initial field that made them move in the first place. At that point there is no longer a field. There is no further force on the electrons, and they stay where they are. The field within the conductor is gone; it has been reduced to zero by the surrounding charges.

"The charges in a conductor," continued Ghardens, puffing significantly on his pipe, "cannot entirely eliminate the electric field close to the surface, because outside the surface there are no mobile charges. Instead, the field falls off rapidly as it enters the metal, just as a short-range interaction falls off away from its source. Once again the presence of other particles has made the massless photons of the electric field behave as though they did have mass.

"But enough of such theorizing," he said abruptly. "The game is afoot and we should away!" In a flurry of activity he rose from his sofa, stripped off his dressing gown, and donned a heavy cape and deerstalker hat. Before Dorothy and her companions had quite realized what was happening, they had been ushered from the house and into a conveniently waiting cab. All, that is, except the Lion. As usual he was too large to fit inside and ran along beside the vehicle, where his presence gave the cab horse all the incentive necessary to proceed at a smart pace. Soon they were leaving the cobbled streets of the town, and a little time later, they found themselves trotting along a rough road, through an increasingly desolate moor.

"The apparent mass that surrounding matter may give to photons is somewhat superficial," remarked Ghardens as they jolted along the track. "A so-called 'real' photon requires little energy to create it and still has no rest mass. It may readily enough shake off the effects of any surrounding charges once it leaves the region in which such charges are present. Particles such as the W and Z bosons of the weak interaction have a more intimate relationship with the particles—the Higgs bosons—that give *them* mass. From this relationship there is *no* escape wherever they may go. These Higgs bosons are not present only within a limited volume of some material. They are present in the vacuum, and that means everywhere, even *within* any material. When I say 'the vacuum,' I mean all of space, whether or not it happens to be occupied by the few trifling particles that compose your solid matter."

The cab stopped in the middle of what appeared to be a dreary wilderness. Ghardens opened the door and swung down to the ground with a burst of enthusiasm, gesturing to the others that they should accompany him. "Here we are at Masskerville Moor," he continued without pause as he alighted on the ground. "As you will be aware, particles in general may exist briefly in the vacuum in a virtual state. They may borrow the energy they need for their creation from the energy fluctuations of the quantum universe. The Higgs bosons are different. They are bosons, and no conservation laws inhibit their production. Furthermore, they do not need to borrow energy. They interact with one another in such a way that the energy involved in their interaction is negative. This means that their very presence lowers the energy of the vacuum, and the universe looks favorably on low energy. As a consequence, the Higgs is always with us.

"The presence of the Higgs boson is thus a universal presence, a given fact in any situation. Its effect is part and parcel of the nature of the particles with which it interacts. When the interaction with the ever-present Higgs causes a W or a Z boson to seem like a massive particle, it produces *all* the attendant consequences of that appearance. The Higgs interaction will cause

R Gilmore

virtual W bosons that are intrinsically massless to produce short-range forces, as you would expect for massive particles. It goes further than this, however. You can create a W boson, converting it from virtual to 'real.' When you do this, the energy that you must give to the Higgs field will be the same as the rest-mass energy that you would expect for a boson that would produce the observed short range of the interaction. Thus the Higgs interaction again makes a massless boson look just like one with mass of its own.

"Even when the W or Z boson has been liberated and becomes a 'real' particle, it is not free of interaction with the Higgs field. This interaction causes it to be sluggish in acceleration when a force is applied, as though it truly had inertia. The particle has at all times to plow its way through a field composed of Higgs particles. This is dragging on its coattails and pulling it back. It is as though it had to move through a bath of thick syrup at all times. In this way also, it seems to have mass."

THE HIGGS MECHANISM

This is an ingenious way to circumvent a couple of uncomfortable facts.

1. There is only one way that anyone has come up with for preventing rampant in-finities when virtual boson amplitudes add. This is the requirement of local phase independence, or "gauge symmetry." The electromagnetic field is al-ready known to satisfy this condition.
2. Gauge symmetry says that the virtual exchange bosons must have no rest mass. The weak interaction has a very short range, and this would imply a *massive* boson.

So far, no free Higgs bosons have been seen, but they could be quite heavy and need a lot of energy to release them.

Everyone is searching for them.

"Wouldn't it be simpler just to say that the W boson *does* have mass and leave it at that?" interrupted Dorothy. "This all sounds implausible. It is far too complicated."

"That is not an option. As you must have been told in the Plaza of the Im-mortals, the exchange quanta of fields have to show 'local gauge indepen-dence' as the only way to avoid the spread of uncontrolled infinities. If you think my recent remarks were complicated, you can have no concept of how complicated *that* would be. The notion of the Higgs, though it may appear unfamiliar and indeed strange to you, is the only option available that seems to agree with everything else we know. When we have eliminated theories inconsistent with previous knowledge, whatever remains, however implausible it may seem, is likely to be the truth," finished Ghardens pontifically.

Then he led his companions to a point that overlooked a great expanse of completely flat and suspiciously innocent-looking ground. "I have brought you to a suitable allegorical landscape to illustrate the effect of the Higgs bosons," he announced. "That is the Boson Bog. It abounds with unobtrusive Higgs that lie in wait to entangle any unwary particles. In your actuality, the Higgs are everywhere and all particles are already entangled, but here you may see things more metaphorically."

"Is it a sort of quicksand then?" asked Dorothy.

"There may be a superficial resemblance, but this is composed of Higgs rather than sand, and its principal role is to be slow, rather than quick. Observe!"

As they watched, a flock of particles came darting across the moor. They flitted to and fro with incredible speed, light-footed and apparently massless. In their carefree play, they strayed unwittingly onto the dire region of the Boson Bog. Immediately they were trapped! The Higgs interactions tugged on them and dragged them down into a sluggish massiveness. Never again would they be able to frolic lightly. They were now bound by the too solid weight of their ponderous nature.

Dorothy and her companions looked on sadly as the particles struggled in the eternal embrace of the Boson Bog. Their shrill cries were piteous to hear. "Oh, the poor things!" exclaimed Dorothy. "Isn't there anything we can do for them?"

"No, there is nothing," returned Ghardens. "It is Nature's way and quite beyond our control. We all have our burdens to carry, and they must carry the burden of their own mass. But now Jouleson and I must leave you," he said abruptly. "Our services are needed elsewhere, and I can see that but a little way across the Moor there is someone you should meet." He indicated a small group a little way off, waved them in that direction, and sprang back into the cab with his faithful comrade. The cab rattled off quickly, bound for his next display of keen observation and incisive analysis. Dorothy and her associates looked at one another, surveyed the distant group, shrugged wordlessly, and began to walk toward them.

As they came close, a splendidly uniformed person accosted them. He was waving an ornate hilt from which protruded not a steel blade but a piece of chalk.

"Hello there!" he cried. "Have you come to inspect our maneuvers? I am General Theory and this is the Standard Model Army." He indicated a group of thirty or so particles that were gathered around him. "A finer body of particles would be hard to find. In fact, it has been impossible to find any other particles at all. In encounters with the ranks of experimental evidence, they have never yet been defeated." Here General Theory leaned down confidentially. "They soon knock the blighters for six, don't you know."

The general drew himself up tall again. "Impressive looking bunch aren't they?" he continued. "Here are our fiercest fermion fighters. On the left is the lepton light infantry, with its three generations of charged leptons and neutrinos. The tau lepton isn't as light as he might be, but we try not to comment," he whispered as an aside. "Then you have the three color regiments: the Reds, the Greens, and the Blues. Each contains six quarks. We also have our boson artillery, which can provide a pretty brisk exchange of photons, gluons, W and Z bosons, and even gravitons, as you may imagine. There is also the quartermaster's department, with the Higgs to provide mass and a selection of appropriate coupling constants to lay down the rules and strengths of engagement. The whole assembly *is* a bit large and unwieldy, but it seems to work."

THE STANDARD MODEL

The Standard Model is a summary of what is known and believed about elementary particles and their interactions. It gives a sort of tool kit for building the matter in our world. Its contents are a number of basic fermions: six leptons and six quarks (the latter in three colors). To bind these together there is a selection of bosons to provide the interactions: the photon, the gluon, the graviton, and the W and Z bosons. Together with a selection of coupling constants and a few phase factors, this collection corresponds amazingly well to the observed results of experiment in particle physics and predicts these results with remarkable confidence and success.

Theoreticians would like to find some more basic theory that would tie together the motley assortment of masses and interaction strengths that appear in the model.

Experimentalists would like, for their own satisfaction, to make a measurement that the Standard Model cannot explain.

So far both groups have been fairly unsuccessful.

"How can you call this a large army?" protested Dorothy in disbelief. "I do not see how you can say it is an army at all. There are only a handful of them."

The general looked quite affronted. "I think that you are confusing variety with sheer number," he said. "When I speak of 'the electron,' for example, I speak of the common nature of all electrons. Particles of any given type, such as electrons, are not just similar; they are *identical*. This identity gives rise to exchange symmetry and the Pauli Principle. Because electrons cannot be distinguished from one another, there is no such thing as a *different electron*. They are all the same, absolutely and identically the same, and so we speak of *the* electron.

"As far as crude numbers go, though, you will find that they are sufficient. Our legions are legion. Observe a sample of our drill if you think that this army is inadequate."

On his command, a small group of quarks stepped forward and began to wheel around one another in a convoluted pattern. Dorothy could see that they had assembled into protons and neutrons. The electron stepped smartly one pace forward from the ranks of the leptons and joined in the weaving march. Now it could be seen that the particles had combined in various ways, and clearly in considerable number, to form whole atoms. There were atoms of hydrogen, of oxygen and nitrogen, of carbon of course, and even of iron and other rarer metals.

This assembly of atoms now converged on one another in a great rushing whirlwind of motion and multiplied in incalculable numbers. The atoms joined together in molecules; they assembled in crystals and amorphous solids. The variety and scope of combination dazzled the mind and was quite impossible for the onlookers to follow. When the blur of activity stilled at last, Dorothy and her friends found themselves confronted by figures clad in dark steel, not one or two but in close order, like the terra cotta warriors of Xian. These were not terra cotta. They were heavily armored, with great razor-sharp swords and other weapons distributed about their persons (if they *were* persons, for the eyes that peered from the thin slits in their visors looked strangely inhuman). The press of great, black-armored shapes darkened the whole plain, extending to the horizon and beyond.

THE EXPERIMENTAL DESERT

Science, and certainly particle physics, depends on experimental results. We cannot determine the nature of the universe by sheer thought. Much of what we have so far discovered we would never have imagined. Current theories of the basic particles suggest that any fresh evidence may come only at much higher energies—energies that would require an accelerator with the diameter of the solar system.

This may be difficult to fund.

"Now perhaps you may agree that this army is adequate for most occasions. Its components are, however, rather worryingly diverse. Why do they exhibit the strange assortment of masses that they possess? Perhaps these are provided by some sort of Higgs process, but if so, why do they have the strange assortment of values that we observe? Why are there three generations of fermions, no more and no less? It all seems strangely arbitrary, and we look for some more basic principles behind this collection that may explain the diversity. So far it has been difficult to find any hint from experiment that a new order is needed. The Standard Model continues to carry all before it. If new observations are to be made, they must be much farther away in realms of higher energy than have been explored so far. Any new results must be a long way into the Great Experimental Desert—that inhospitable region of ever-increasing energy where new observations may lurk—but as yet there is no sign that any are within reach of a conceivable experimental expedition. My fellow officer, Major Discovery, has set off to reconnoiter the region, but it may be a long time before he returns with news. If ever," he added thoughtfully.

"I have been told that I must pass through the desert, so perhaps I may find the major," suggested Dorothy helpfully. The general did not seem much encouraged, so they left him still looking pensive and set off toward the line of sand on the horizon that marked the beginning of the Great Experimental Desert.

CHAPTER 13

THROUGH THE GREAT EXPERIMENTAL DESERT

The companions had scarcely begun to cross the desert before they felt oppressed by the vast extent ahead of them. Far into the distance stretched the barren sand, all detail masked by the fierce haze of ever higher and higher energy. Each step of the way became successively more difficult as they realized that they had somehow to achieve the almost unattainable energy represented by the way ahead. Their surroundings shimmered with greater and greater virtual activity as energy levels generally increased. The environment played strange tricks with their eyes.

"In this place I am not sure that I would trust what I see," remarked the Scarecrow. It was not easy for him to confess that even his powers of observation might not be equal to the task. "We may well experience mirages."

This remark caused everyone to look around, and indeed they did observe a patch of green against the yellow sand some way ahead. As they trudged closer, this resolved into a pleasant patch of grass, with a pool in the middle and a few trees to the side. Beneath the largest tree reclined a young man clad in silk robes and wearing a silken headdress to give him further shade. By his side rested a loaf of bread and a jug of wine, from which he refreshed himself at intervals. On either hand he had a row of companions, two sets of particles that flanked him. To the right they could see the familiar set of fermions and bosons: leptons and quarks, photons and gluons, and W and

Z bosons. To the left was the set of supersymmetric partners. Again there were fermions and bosons: photinos, gluinos, winos, sleptons, and squarks. The travelers could see that this personage was indeed Supersymmetry, now come to his full manhood (theoryhood?). He lay back with a faint expression of smugness and took his ease in this oasis of experimental justification.

"I am not sure I trust this," whispered the Scarecrow. "I think it may be our first mirage." Already the scene of blissful satisfaction was beginning to look a bit tenuous around the edges. Dorothy and her companions turned their backs on it, gritted their teeth (those who were equipped with such things), and resumed their slow progress toward the distant horizon.

They could in no way judge how far they had gone when once again a distant feature became visible. Eventually it could be seen quite clearly. As they approached, already dazed by their long trek across the Great Experimental Desert, they seemed to hear a voice declaiming within their heads.

> *A vast and footless TOE[1] stood in the desert. Near it on the sand,*
> *half sunk a shatter'd GUT,[2] full of the scraps of half-digested theories.*
> *These tell their authors well their concepts wrought,*
> *Explaining all and yet predicting naught.*
> *And on the pedestal these words appear: "Behold the final theory, first*
> *and last. Look on my simple beauty and rejoice!"*
> *Before the TOE the theorists lie prostrate,*
> *Worshipping this vision of their fate.*
> *Nothing else remains. While other theories fall into decay*
> *And yield their content to the mighty TOE, all round about*
> *The lone and level sands stretch far away.*

Dorothy and her fellows looked at one another.
"It's best not to even think about it!" said the Lion.
"Pretend you didn't observe it," said the Scarecrow.
"It is beyond my processing capacity," said the Tin Geek.

---

[1] Theory Of Everything. This is the Holy Grail of the unified theories. It would be a theory that unites in one grand hypothesis all the interactions, *including gravity*. This is not easy to achieve.

[2] Grand Unified Theory. There are a lot of these. Their objective is to include all the particle interactions, but possibly excluding gravity. The inclusion of gravity would yield a Theory of Everything (TOE).

R Gilmore

Without even stopping to investigate, they continued on their journey, leaving the stark and enigmatic sight behind. On and on they traveled, past all hope and perhaps past all reason, as the Wizard had previously warned them. Step after step, their energy increased with the gradualness of an eternity. But even this seemingly endless desert eventually had its end, and they found themselves at a new and different region. They were again on a regular track and a sign by the side read:

Congratulations!
You have reached the Planck Energy.
Your elevation is
10,000,000,000,000,000,000,000 MeV
(approximately)

"I never thought to see this," breathed the Lion, his usual ebullience subdued by awe. "This is the ultimate limit. It is here that the different interactions blend together and become of equal strength. This is where. . . ."

"**Gravity becomes strong**," thundered a voice that seemed to come from all around them, to the left and to the right, but particularly from overhead. They looked up and saw the Witch of Mass, but now she filled the whole field of vision. In her presence there was nothing else *to* see.

GRAVITY

Of all the known interactions, only gravity has so far resisted attempts to create a satisfactory quantum theory. Gravity is different from the other interactions in that gravity distorts space and couples directly to energy. This means that as particle energies increase, gravity becomes stronger.

At any normal energy, gravity is extremely weak. But at high enough energies, gravity will be strong, and then *nothing* can resist it, because resistance involves some form of repulsion, and repulsion builds up energy. This energy only makes gravity stronger, so eventually, any resistance to the pull of gravity will in fact *assist* gravity. This irresistible cycle is seen in the formation of a black hole, where none of the other forces brought to bear on matter can resist effectively, and gravity crushes whole stars—and even the hearts of galaxies—down to a point in space.

"Now you see my full power," she said more moderately, though still with overwhelming power. "For I am different from any of my three sisters. Each of their interactions has its specific charges—electric, color, or whatever. These charges fix finally and absolutely the forces that their interactions may exert. My interaction couples directly to mass, and mass is energy. At low energies I am weak, but here, where the energy levels are huge, I am supreme. Resistance to me is useless. Actually, it is worse than useless, for any attempt to resist my attraction must involve repulsion and its attendant energy. Such energy only increases the force of my attraction. Nothing can stand against me. Even space itself is unable to withstand the impact of my presence."

The Witch of Mass then raised her fist and brought it down upon a region of empty space. At the point where she had struck, the very fabric of space was shattered and distorted, ceasing to be a placid background for the activities of matter. Space itself bubbled into a sort of foam, and wormholes opened within it. This cataclysmic disruption threw Dorothy and her companions violently about, and the girl found she was approaching the writhing entrance to a wormhole. Even in this extremity, she was somewhat surprised to note that at the entrance to the wormhole, there was a subway sign. She had no time to consider this as she shouted goodbye to her companions and tumbled inside.

She found herself within a sort of tunnel and, looking back, could see one of the Witch's eyes peering down the tunnel after her. Then the Witch winked at her, and as the giant eye closed, so did the entrance to the worm-

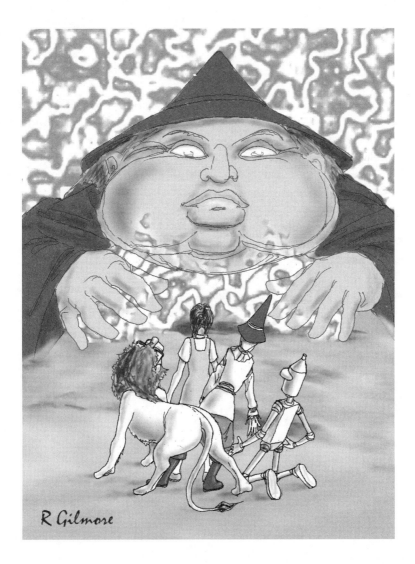

R Gilmore

hole, sealing Dorothy off from the external universe. There was nothing to do but to go on farther down the passage in which she found herself.

Soon she came to a barrier at which stood a uniformed official.

"Ticket please!" he said.

"I'm sorry," said Dorothy. "What ticket do you mean?"

"Why, your subway ticket, of course. I hope you haven't lost it."

Dorothy searched through her pockets and found that indeed she still had the ticket she had purchased when she set off with her aunt and uncle. She held it out, and the ticket collector punched it.

"Is this really part of the subway?" she asked, finding it hard to believe.

"Oh certainly. Most people are surprised to discover just how far the subway extends beyond the city center. I hope you have learned an important lesson on your trip."

"Oh yes. I have learned how the everyday world I live in is founded on the strange and marvelous behavior of countless tiny particles and their amazing interactions."

"Well, yes. I was thinking more of the lesson 'Don't lose your subway ticket,' but that other is certainly a most important lesson. Anyway, have a nice journey."

Dorothy went on down to the platform, at which she found an empty subway car waiting for her. She entered it and found it filled with that half melancholy feeling of loneliness that haunts carriages when they reach a distant terminus. She chose a seat, sat down, and waited.

The subway car gave a sudden lurch. . . .

INDEX

The index does not list every occurrence of a word. It shows where concepts are first introduced or explained.